Report Writing for Code Inspectors

Jean Reynolds, Ph.D.
Polk State College
Winter Haven, Florida
and
David Diamantes
Code Trainer and Consultant
Berryville, Virginia

The Maple Leaf Press

Published by CreateSpace

www.InspectorWriteRight.blogspot.com

Printed in the USA

Copyright © 2014 Jean Reynolds and David Diamantes

All rights reserved.

Table of Contents

Introduction   3
Pretest   5
Pretest Answer Key   8
Section I: Writing the Report   11
    Chapter 1 Why Is Report Writing Important?   13
    Chapter 2 Introduction to Report Writing   19
    Chapter 3 Preparing to Write   27
    Chapter 4 Organizing a Report   33
    Chapter 5 Writing a Report   35
    Chapter 6 Reports, Notices, Citations, Summonses and Orders   41
    Chapter 7 Objectivity   47
    Chapter 8 What To Omit   51
    Chapter 9 Quoting Exact Words   55
    Chapter 10 How Helpful is OJT?   59
    Chapter 11 Bullet Style   63
    Chapter 12 Active or Passive Voice?   67
    Chapter 13 Online Resources for Code Inspectors   71
Section II: Solving Sentence Problems   75
    Chapter 14 Fragments   77
    Chapter 15 Run-on Sentences   81
    Chapter 16 Misplaced Modifiers   85
    Chapter 17 Parallelism   89
Section III: Professional Sentences   93
    Chapter 18 Writing Effective Sentences   95
    Chapter 19 Periods and Semicolons   97

Chapter 20 Commas	103
Chapter 21 Comma Rule 1	105
Chapter 22 Comma Rule 2	109
Chapter 23 Comma Rule 3	113

Section IV: Mastering English Usage — 117

Chapter 24 Apostrophes	119
Chapter 25 Quotation Marks	125
Chapter 26 Pronouns	127
Chapter 27 Verbs	133
Chapter 28 Subject-Verb Agreement	139
Chapter 29 Capital Letters	143
Chapter 30 Comparisons	147
Chapter 31 Prepositions	151
Chapter 32 Avoiding Common Errors	155
Chapter 33 Myths about Grammar	161

Section V Choosing the Correct Word — 165

Chapter 34 Criminal Justice Terminology	167
Chapter 35 Words and Expressions to Avoid	171
Chapter 36 More Words to Watch	177
Chapter 37 Using Plain English	187
Chapter 38 The Freedom of Information Act	193

Post-test	197
Post-test Answer Key	199
Answer Key	202
Index	227
About the Authors	229

Introduction

Effective writing is vital to your career as an inspector. Your notices of violation, letters, and reports are public documents that may be read by supervisors, attorneys, judges, citizens, design professionals, contractors, and reporters. Quality inspection reports and notices of violation help ensure code compliance, impress superiors, and win respect from colleagues. They facilitate repairs, reinforce requirements for safe operations, and ensure unsafe practices are discontinued.

If you work for a government agency, your notices, letters, and inspection reports may become the basis for appeal hearings and criminal and civil trials—and may even prevent cases from ending up in court. Attorneys, consulting engineers, and code consultants who read your reports hoping to find omissions and errors may decide that the odds of prevailing in an appeal or trial are slim, and compliance is their best option.

Unlike police reports that capture the details of an incident or complaint for further investigation or prosecution, your inspection reports and notices of violation will compel a business or property owner to take action to bring the property or facility into compliance with the code. Clarity is important because repairs and changes to procedures and practices generally involve precious time and money.

Your writing skills can help you advance in your career. You'll be prepared to communicate effectively with business owners, design professionals, the media, community leaders, and other government officials. Well-written documents help you create a reputation for professionalism, accuracy, and fairness that will stand you in good stead as you start to climb the career ladder.

This book offers you a wealth of information about report writing. A pretest will help you assess your strengths and determine which skills need your attention. Section I shows you how to organize and write professional reports. Sections II and III cover sentence skills, Section IV helps you avoid usage errors, and Section V covers special words you need to know. A post-test helps you decide what areas need further review. Exercises are provided throughout the book, and an Answer Key allows you to check your progress at each step. Let's get started!

Acknowledgments

from Jean Reynolds: To my husband, Charles Reynolds, for his long and loving support through all my writing projects, including this one. And of course to my co-author, David Diamantes, a great collaborator who brought a wealth of code enforcement experience and knowledge to this book and—indeed—suggested this project in the first place.

from David Diamantes: To Jean Reynolds for taking me under her wing, providing the brains, and doing the heavy lifting on this project, and to my wife Bonita for always supporting my efforts.

PRETEST

Instructions: Complete each activity below. When you're finished, check your responses against the Pretest Answer Key beginning on page 2.

Part I Effective Reports

Instructions: Put a check √ if a sentence meets the requirements for an effective report. Put an X if the sentence does not meet the requirements.

____1. The car was parked within three feet of the fireplug.

____2. Ms. Wright was belligerent when I arrived for the inspection.

____3. I looked through the open front door and saw a man with both his hands around a woman's neck.

____4. Carol Todd was watching television while I interviewed her husband.

____5. I saw Fowler's car cross the double line three times as he drove down Second Street.

____6. It was obvious that Fowler's driving was impaired, probably by alcohol.

____7. I asked Ms. Barker when she last heard from her landlord.

____8. Ms. Wright refused to cooperate.

Part II Effective Word Choices

Instructions: Put a check √ if the wording of the sentence meets the requirements for a modern report. Put an X if it does not.

____1. Sarah Wilson advised me that she had left for work at 7:45 that morning.

____2. A leaking kitchen sink was found when Ms. Wright's apartment was inspected by this inspector.

____3. I ascertained that Buckley was the tenant.

____4. I asked Anderson if she needed medical help, and she answered in the affirmative.

____5. Sawyer cursed and swore when he saw my identification.

____6. Lillian Thompson said she planned on flipping the house.

____7. The window over the kitchen sink was broken, and pieces of glass were lying on the kitchen counter.

____8. When I contacted Jeffrey Klein, the neighbor, he denied having seen or heard anything unusual.

Part III English Usage

Instructions: Put a check √ in front of any sentence that meets the requirements for English usage. Put an X in front of any sentence that does not meet the requirements.

____1. Paula Dillon gave permission for my partner and I to inspect her apartment.

____2. The Quinns have lived in the house on Central Boulevard for two years.

____3. The women's descriptions of their attacker were detailed and consistent.

____4. There's records of two previous complaints for overgrown vegetation at that address.

____5. Felicia Jones told me that her Father had been gone all weekend.

____6. Mr. Thompson told me that "I wouldn't find any violations in his restaurant and should leave him alone."

____7. The cut looked serious, it obviously needed medical attention.

____8. The neighbor whom I interviewed gave me a description of the suspect.

PRETEST ANSWER KEY

Page numbers refer to sections of this book with information about the skills on this test.

Part I Effective Reports

X 1. The car was parked within three feet of the fireplug. [*hydrant* is the proper technical term used in the code.]

X 2. Ms. Wright was belligerent when I arrived for the inspection. [Page 56: Vague. State exactly what Ms. Wright said or did.]

√ 3. I looked through the open front door and saw a man with both his hands around a woman's neck.

√ 4. Carol Sanders was watching television while I interviewed her husband.

√ 5. I saw Fowler's car cross the double line three times as he drove down Second Street.

X 6. It was obvious that Fowler's driving was impaired, probably by alcohol. [Page 56: This kind of statement may not fare well in a courtroom. State what you saw Fowler do, and omit your opinion.]

√ 7. I asked Ms. Barker when she last heard from her landlord.

X 8. Ms. Wright refused to cooperate. [Page 56: Vague. State exactly what Powers said or did.]

Part II Effective Word Choices

X 1. Sarah Wilson advised me that she had left for work at 7:45 that morning. [Page 139: Sarah Wilson *told* you she had left for work. Save "advise" for actual advice.]

X 2. A leaking kitchen sink was found when Ms. Wright's apartment was inspected by this inspector. [Page 71: Avoid passive voice. A better version would be: "I found a leaking kitchen sink when I inspected Ms. Wright's apartment." See page 22 for more about using "I" and "me" instead of the outdated phrase "this officer."]

X 3. I ascertained that Buckley was the tenant. [Page 171: "Ascertained" is jargon that should be avoided. Another problem: This sentence doesn't explain how you got this information.]

X 4. I asked Anderson if she needed medical help, and she answered in the affirmative. [Page 174: Say simply that *she said yes* or *nodded her head up and down.*]

X 5. Sawyer cursed and swore when he saw my identification. [Page 179: "Curse" means to call down evil powers; "swear" means taking an oath. And you should write exactly what Sawyer said, word-for-word, instead of generalizing: See page 56.]

X 6. Lillian Thompson said she planned on flipping the house. [*Flip* is a slang term for reselling for profit.]

√ 7. The window over the kitchen sink was broken, and pieces of glass were lying on the kitchen counter.

X 8. When I contacted Jeffrey Klein, the neighbor, he denied having seen or heard anything unusual. [Page 174: "Contacted" is vague: Did you phone, visit, or email Klein?]

Part III English Usage

X 1. Paula Dillon gave permission for my partner and I to inspect her apartment. [Page 129: *my partner and me*]

√ 2. The Quinns have lived in the house on Central Boulevard for two years.

√ 3. The women's descriptions of their attacker were detailed and consistent.

X 4. There's records of two previous complaints of overgrown vegetation at that address. [Page 139: *There are records*]

X 5. Felicia Jones told me that her Father had been gone all weekend. [Page 144: Lower-case *father*]

X 6. Mr. Thompson told me that "I wouldn't find any violations in restaurant and should leave him alone." [Page 125: Not Thompson's exact words, so delete the quotation marks.]

X 7. The cut looked serious, it obviously needed medical attention. [Page 81: There are two sentences; change the comma to a period, and capitalize *it*. You can also use a semicolon and leave *it* lower-case.]

√ 8. The neighbor whom I interviewed gave me a description of the suspect.

Section I: Writing the Report

Chapter 1

Why Is Report Writing Important?

Think back to when you first decided on a code enforcement career. What attracted you? Chances are it *wasn't* report writing. Inspectors frequently say that writing is *not* a favorite task: It's time consuming, tiring, and exacting—and there are serious consequences if they make a mistake.

But report writing is essential to your career as a code professional, and writing becomes even more important as you advance up the career ladder. In fact it can even help your superiors decide that you're qualified for promotions and greater responsibility.

Your reports are the catalysts that cause unsafe conditions and violations to be corrected. Just identifying a problem isn't enough. In most cases, compliance with improperly written notices or orders is not mandatory. By definition, notices of violation that lack key elements, such as code section and time limits for correction, are not legal notices at all.

First Impressions

Don't judge a book by its cover. You've heard it a thousand times, and like most people, you probably ignore this sage advice at times. But evidence of the importance of first impressions is all around you. Publishers invest millions of dollars on slick covers that grab your attention as you stare at the rack of paperbacks in the airport newsstand. Stores spend vast amounts of money on window displays designed to lure you inside.

First impressions matter. You get only one chance. Business owners, design professionals, citizens, and other public officials will make an unconscious snap judgment the moment they meet you for the first time. In the inspection business, that first impression will often involve your report rather than your person.

In the wake of the 2003 Station Nightclub fire in which 100 people were killed, copies of previous fire inspections reports were published in newspapers and online. Reporters, attorneys, members of the grand jury, and the public got their first impression of West Warwick, Rhode Island's fire inspector, as they looked at those images. Those readers were likely to have little technical expertise in fire and building codes, so they evaluated the fire inspector's competence based on grammar, penmanship, and completeness.

Setting the Record Straight

Is there ever a time you should write a report, even though your agency policy doesn't require one? Consider this case from Reading, Pennsylvania. In *Mary Ann Ciarlone v. City of Reading*, a landlord and her three tenants alleged that the City of Reading violated their Fourth Amendment and Fourteenth Amendment rights when city code inspectors used a sledgehammer to execute an administrative search warrant. The case presented the novel legal question of whether force may be used to execute an administrative search warrant to conduct a routine property maintenance inspection.

A friend of Ms. Ciarlone videotaped the warrant service and posted it on YouTube. In addition to alleging that the warrant was not supported by probable cause and was unreasonably executed, the plaintiffs also contended that the City violated their Fourteenth Amendment substantive due process rights because the City allegedly retaliated against Ms. Ciarlone and singled out the property for the use of a sledgehammer. The inspector's written reports were key evidence in Federal Court.

After a two-week trial, the jury found in favor of the defendants (City employees).

In the Ciarlone case, the inspectors knew there would be repercussions, and a lawsuit was possible. Ms. Ciarlone told them up front to get the warrant: She would not willingly allow them to enter her property. As in this case, there will be inspections where your instincts tell you *there is more to come*. Follow your instincts. Record the details in a report while they are fresh in your mind, even if you are not directed to do so. Write a narrative report for the file. Write down exactly everything that occurred and everything that was said. It could be the best career decision you ever make.

What Goes into a Good Report?

Report writing can sound intimidating if you're new to code enforcement. It's important to know, however, that several factors are working to your advantage. First, you're already a writer. The writing skills you learned in school will give you a good foundation to build on as you learn about report writing.

Second, help is available if you need to brush up on grammar and usage. Here's a summary of the skills you need to write effective sentences, paragraphs, and reports:

- three comma rules (page 103)
- two ways to use apostrophes (page 119)
- four pronoun rules (page 127)
- five rules for capital letters (page 145)
- four subject-verb agreement rules (page 139)

In addition, you need to master technical code vocabulary, and you need to watch for some commonly misused words (*your/you're*, *to/too*, break/brake, and others—see pages 179-88).

And here's a tip: The chapters about Avoiding Common Errors (page 157) and Myths about Grammar (page 163) are short and readable, and they can teach you a lot about writing in a short time.

All the writing skills you need are covered in this book, and practice exercises and answers are included. You can also ask family members, friends, and co-workers to look at your writing and help you spot problem sentences. Take note: With practice, *every inspector* can become a competent writer.

Third—and this is perhaps the best news—report writing is predictable, even though code inspections work is not. Most code inspectors write only a few types of reports. Learn the basic requirements for report writing, and you're assured of producing an effective report every time you sit down to write.

The bottom line is that professional writing skills are within reach of *any inspector*—including you—provided, of course, that you're willing to invest the time and energy needed to be an effective writer.

Exercise 1 Why Are Reports Important?

Instructions: Imagine that a friend has been talking with you about a possible career in code enforcement. He or she is looking forward to working in the field. Your friend disliked English in high school and hopes to spend as little time as possible writing reports on the job.

Write a short letter explaining why report writing is important and offering suggestions for sharpening the skills needed. When you're finished, check your ideas against the list on page 202 in the Answer Key.

Chapter 2

Introduction to Report Writing

Report and notice of violation formats vary from locality to locality and agency to agency. Some inspectors write their inspection reports on blank sheets of paper, while others use paper forms or computer templates with boxes and bubbles ready to fill in. But *all* reports share some common features, and all require the same qualities:

- accuracy, brevity, and completeness
- objectivity
- a clear description of the unsafe condition or violation
- the required corrective action(s)
- elements such as the right to appeal, code or ordinance section, reinspection date or other information required by the code or by law

Although details may vary, depending on the agency or institution you're working for and the kind of situation you're dealing with, the basic features of a report are always the same. You begin by establishing the date, location, name of the business, name of the owner or representative, and type of inspection.

Inspectors in agencies that use tablet computers or personal data assistants (PDAs) with pick lists and checkboxes might wonder why writing is a necessary skill. Even inspectors whose systems are totally digital are still required to write narrative reports on occasion. Enforcement actions, court cases, and appeals hearings normally require a narrative report to describe the property conditions or violations, the inspector's actions and orders, and the property owner's actions or inaction. Narrative reports are also required when there is an accident or a personnel action. If you are interested in promotion, an assessment of your writing skill will likely be part of the promotion assessment process.

Sample Reports

Let's examine two reports—one for a property maintenance complaint, and one for a fire inspection bureau—to see how they're organized.

Report for a Property Maintenance Code Department

This report was written by Inspector Carole Donner when she was notified of an unsecured fire damaged property.

> At 3:20 p.m. on October 3, 2012, I, Inspector Carole Donner, was notified by radio that the house at 35 Woodland Road was vacant and unsecured. The house was damaged by fire and had been boarded up, pending repair.
>
> I talked to Sam Farley, the next-door neighbor, who resides at 33 Woodland Road. He told me that he had seen teenagers loitering around the house for the past several days. He noticed that plywood was missing from a broken window on the west side of the house when he walked his dog this morning.
>
> I walked around the perimeter of the house and saw and heard no one. A piece of plywood the approximate size of the window was leaning against the exterior wall near the window. I shone my flashlight in the window and saw no evidence of entry. Tax records indicate the owners are Margaret and James Rice. Mr. Farley gave me James Rice's cell phone number (717-XXX-3393). I called Mr. Rice and informed him of the unguarded window. He is a contractor and will repair the fire damage when he receives the insurance

settlement. Mr. Rice said he will secure the window this afternoon. I will re-inspect tomorrow, October 4, 2012.

Report for a Fire Prevention Bureau

This report was written by fire inspector Frank Dunham.

> At 10:35 a.m. on July 9, 2010, I, Inspector Frank Dunham, was conducting a fire inspection at Bridges Auto Body at 5598 West Locust Street. While inspecting the rear of the building from alley, I saw the fuel oil tank for the building directly behind Bridges (Taiwan Garden) had a large dent and appeared to have been struck by a vehicle. There was no evidence of a leak. The tank appeared to be new and did not have required vehicle impact protection installed.
>
> After finishing my inspection of Bridges Auto Body, I entered Taiwan Garden (5597 West Maple Street) and spoke with owner Harry Cho. I asked Cho if the tank was new. He stated the original tank in the basement had begun to leak, so he had the new tank installed. I asked if his heating contractor had secured a permit to install it. He said he didn't know and gave me the name and number for Skip's Mechanical (515-XXX-2132).
>
> I called Skip and told him he needed to secure a mechanical permit from Permits and Licensing and provide vehicle impact protection within 24 hours. He said I didn't have the authority to require him to get a permit. I told him the tank had already been struck and he would be responsible for cleanup if there is a spill. He agreed to get a permit and install bollards around the tank.
>
> I called Brenda at the permit counter and informed her of my actions. I will follow-up with the Mechanical Department and re-inspect tomorrow, July 10, 2012.

A Closer Look

Take a moment to reread these two reports. Then list the features the reports have in common. When you're finished, compare your list to the list below.

Here are some of the features you might have noticed:

- The reports have a beginning, middle, and end
- Both inspectors explained why they were at the scene
- The date, time, and place were recorded, along with names of persons they talked to
- The inspectors recorded what they saw and did, along with what they were told
- The reports ended with a wrap-up explaining the outcome of the situation

You might also have noticed that the inspectors used "I" and "me" when referring to themselves. They wrote clear, simple sentences, and they stuck to the facts, avoiding opinions and guesses about what had happened.

The Stages of Report Writing

Like any writing task, report writing proceeds in three stages: Preparation, drafting, and revising.

Preparation includes observing, interviewing, investigating, and taking notes.

Drafting involves organizing and recording the information on paper or a laptop. You may be given a paper form with spaces for names, date, location, offense, and other information. If you're using a laptop, you'll be typing this information into spaces on the screen.

Revising includes spellchecking, verifying information, and checking for correct English usage, clarity, completeness, and professional style.

This book will offer you tips for effectively completing every step in the report-writing process.

Meeting the Challenge

Inspectors who are new to the code enforcement field sometimes underestimate the sophisticated thinking skills required for effective report writing. As you write your report, you may need to:

- blend two sets of stories—what happened before you arrived at the scene, and what you observed yourself
- accurately recall and record what business owner and witnesses tell you
- sift through conflicting accounts to determine what really happened
- select the information needed for follow-up inspection or investigation, if necessary
- eliminate bias and emotion from your account
- justify your actions
- clearly state what actions need to be taken to correct the unsafe conditions or violations
- carefully note all details that may be needed for future appeals or trials

Here are some important points to remember when you write a report:

1. Use names.

Avoid labels like "owner" or "tenant," which quickly become confusing. Give the person's full name the first time, and then switch to last names only. If two or more people have the same last name, you can use their first names. Don't use "Mr." and

"Mrs.": Once you've established the family name as "Johnson," you can refer to the spouses by their first names.

2. Be efficient.

Don't write "month of September": September is always a month. Don't write "for the purpose of" when you mean "for" or "to." You can see a list of time-wasting words and expressions beginning on page 173.

3. Don't write statements that might be challenged.

Avoid hunches, guesses, and predictions. Don't say, for example, that a person *attempted*, *tried*, *intended*, or *planned* to do a particular action. Write only what you've seen or heard: You saw the owner spray-painting outside the spray booth, washing machine parts in gasoline or illegally take water from a fire hydrant. (See page 47 to learn more.)

4. Be complete.

If you contacted another agency or department, include the name of the person you contacted and your method of communication. If you took samples for testing, include the results, even if they were negative. If you took evidence away from the scene, list each item and explain what happened to it ("chain of custody"). If you called for assistance from fire, police, health, or another inspection agency, note those actions in your report.

5. Don't generalize.

Avoid vague wording: *I contacted Bill Gates.* (Did you phone him, email him, or see him face-to-face?) *I noticed the property was for sale.* (Did you see a for sale sign in the yard? Did you see a newspaper ad? Did you find the information online?)

6. Don't use industry slang or acronyms that may be misinterpreted.

Avoid terminology that might confuse your readers. For example, *closed down* raises questions about whether you have the code authority to close a business or whether you ordered the evacuation. Civilians may not understand what *FDC* (fire department connection) means, and they probably won't be familiar with *Side A, B, C* or D (use terms like *front, rear*, and *north, south, east*, or *west side* instead). And don't use terms that might have multiple meanings. For example, *A/C* could signify either "air conditioning" or "alternating current."

7. Write like the professional you are.

If you're writing on a computer, use the spellchecker and grammar checker. These electronic tools are not foolproof, but they will catch many errors. Make a list of words that you have difficulty spelling, and make sure it's handy whenever you write a report. Whenever possible, have someone read over your reports before you submit them to a supervisor.

Exercise 2 Rewrite a Paragraph

Instructions: A paragraph from a report is printed below. Using what you have learned, evaluate the paragraph. (Do not be concerned about the parts of the report that have been omitted.) When you're finished, go to page 202 to check your answers.

The property was overgrown with weeds, and trash was accumulated on the carport. I asked Tenant 1 (Maria Hernandez) the name of her landlord but she acted like she didn't understand. Tenant 2 (Luis Ochoa) said they pay the rent directly to a man named Ted, who comes by every Friday afternoon. I asked Tenant 2 for Ted's phone number. He said he didn't know. I asked Tenant 2 to ask Tenant 1 if she knew it. She shook her head.

Chapter 3

Preparing to Write

A good report begins before you start writing. As you're observing, interviewing, and taking notes, you need to make an extra effort to ensure that you have all the facts needed for a complete report. Whether you are documenting an inspection, recording the details of a meeting or interview, or preparing a document for an appeals board or the court system, you need to capture the facts while they are fresh. You probably won't get a second chance. Here are some important guidelines:

1. Be prepared to take notes.

Of course you have writing paper (and perhaps a laptop). But what if you jump out of your car to deal with an emergency? It's embarrassing to be caught without writing materials. Go to the Dollar Store and buy a few tiny notebooks. Keep one in a pocket, along with a couple of pens, just in case you need it. (A reminder: If you keep your notes, they're subject to subpoena and may be subject to your state's Freedom of Information laws. Don't mix personal information with your job-related notes.)

2. Think ahead.

Capture as much information as possible. You probably won't get another chance. The neighbor who offers information about a homeowner's unlicensed repair garage may rethink talking to local officials and refuse to speak with you. Subcontractors on construction jobs may have completed their work and left the area.

3. Think about the type of report or notice you'll be writing.

Notices of violation typically have required elements that must be included. Legal names, property descriptions, dates, and details about equipment or vehicles may be required. An informational report regarding an incident or report of an unsafe condition may include the names and statements of witnesses. You cannot acquire too much information. The informational report may lead to further investigation and the issuance of a notice or citation. It may result in a court case or appeal hearing. Follow your instinct when something in the back of your mind tells you *this situation has legs*. Think ahead.

4. Train yourself to observe and remember.

Make an extra effort to look, listen, and remember, especially when you first arrive at a property or incident. Who met you when you arrived? Why were you there? What was the condition of the property or equipment?

5. Record information promptly and thoroughly.

Don't rely on your memory to add details lately. It's embarrassing to be caught with an inaccurate or incomplete report. Discipline yourself to write a complete set of notes as soon as possible.

This completeness requirement may sound easy: Just write down everything that you saw or heard, right? Unfortunately, it's not that simple. Inspectors often forget to record a piece of essential information. For example, an inspector might mention a defective gas furnace resulting in carbon monoxide in a building but forget to identify the manufacturer and model of the unit. In most cases, codes require inspectors to identify themselves and request permission to enter the premises. If the case ends up in court, expect to be asked who granted that permission. This is particularly important if more than one inspector is at the property. "I thought my partner asked" won't get you far.

Interviews

Talking to owners, tenants, and witnesses can present challenges. Depending on the type of inspection, stress levels may be high, and you may be listening to a jumble of relevant and irrelevant information. Tenant-landlord disputes can evolve into neighborhood squabbles. Employee complaints regarding unsafe conditions may be retaliation for termination and may be met with hostility by management. Sorting everything out and accurately recording what you heard can be a complex and time-consuming task.

1. Deal with emotions first. Don't let things get personal. Unlike a police officer called to a crime scene, in most cases, you have the option to conduct the inspection of interview after a cooling off period. Ask to speak with the business owner in their office. Explain why you're there and that you realize how valuable their time is. Then explain that you need the person's help in order to complete your work and not take any more of their time.

When you're calm and professional, the person who's talking is more likely to cooperate and answer your questions. Don't hesitate to interrupt, gently, if a person goes off on a tangent.

2. Provide as much privacy as you can during the interview. People may be less forthcoming when they're being watched, especially if the person who's making the complaint is at the scene.

3. Remember that hearsay is permissible in reports—and it can provide valuable clues for potential future investigations.

4. Use quotation marks in your notes any time you write down a person's exact words. That information may be useful in a court or appeals hearing later on.

5. Don't rely on your memory. You can't always predict how much time will pass before you get a chance to write up your notes as a formal report.

6. Record slang and bad language, even if it sounds unprofessional. Knowing a person's exact words may be useful

in an investigation or court hearing later on, for example. Be sure to ask for explanations if a witness or suspect uses unusual slang or vague language. Children, too, may have difficulty finding the right word.

Completeness

Here are some tips to ensure your report is complete:

1. Make an extra effort to get contact information from anyone who might assist in a follow-up inspection or investigation. If you suspect you might have difficulty reaching a witness, ask for a backup telephone number for a friend or family member.

2. Always document the results of an inspection or investigation, even if no violations were found or the results were negative.

3. If potential criminal activity is discovered, you are a very important witness. Your testimony may be required for the police to secure a criminal search warrant. Why you were on the scene and who granted you access become even more important. Document everything you saw or heard, and try to preserve the scene and any evidence as you found it.

4. Document any steps you've taken to protect a crime scene.

5. If you take samples or remove defective equipment, be sure to establish a chain of custody for evidence. What items did you remove from the scene, and what actions (tag, package, mark, sketch, log) did you take?

Exercise 3 Preparing to Write

Instructions: Choose the correct answer to each question below. When you're finished, check your answers on page 203.

1. Dealing with a victim's emotions
    a) is not part of an inspector's job
    b) should usually be the first step in an interview
    c) should be done only after all the facts are recorded
    d) is rarely necessary

2. "Chain of custody"
    a) refers to transporting a suspect
    b) refers to filing a report
    c) refers to evidence taken at the scene
    d) does not need to be recorded in a report

3. Having extra paper and pens in a pocket
    a) may be helpful in an emergency
    b) is unprofessional
    c) violates most agency's regulations
    d) may damage an officer's uniform

4. Which of the following does *not* need to be documented in a report?
    a) manufacturer and model of equipment
    b) address and legal description of the property
    c) Owner or owner's representative
    d) the inspector's theories about how the conditions occurred.

5. Slang

a) has no place in a report
b) may require a definition if it's unfamiliar
c) should be used only if it's grammatical
d) should be used only if it's easily understood

Chapter 4

Organizing a Report

Most inspections reports are prepared with fill-in-the-blank forms, either with paper or on a computer. Narrative reports are rarely required. When they are, it's usually because situations or events pose a serious public safety hazard, or have the potential to result in litigation. Multiple agencies typically respond and play a role in mitigating the incident, or providing support. By their very nature, these reports are challenging to organize.

A good report is actually a combination of many groups of information. Much of the report will be a simple narrative (story), listing in order the events that happened. But reports may contain flashbacks, and many have contradictory statements from witnesses who have different versions of what happened. Even experienced inspectors say that organizing all this information can be difficult. Here are some guidelines.

You should start with an introductory sentence that will answer *Who, What, When, Where,* and *Why.* (Note that all this information may not be necessary if you've entered it elsewhere on your laptop.)

> I was dispatched to 325 West Branch Street at 1030 hours, on March 18th, for a report of an unsecured building.

Use "I" because your name is already at the top of the report for the *who*. Include *what, when, where* and *why*. Use a logical sequence, start with your trip to the scene, what you noticed upon arrival, whom you spoke with or saw, and your assessment of the situation. Perhaps the next-door neighbor made the complaint of the unsecured building. She saw teenagers pry plywood off a window and enter. Include her statement and contact information. Use quotation marks when quoting her directly.

Summarize your actions, where you walked, what you saw, heard or smelled. If you took photographs, note that in the report. Finish the report with your findings. A piece of plywood was on the ground next to the unsecured window. You contacted the owner by telephone and issued him an order to secure the building in person.

Every encounter with the public and every incident, regardless how commonplace it may seem at the time, has the potential to become serious. An unsecured building might seem trivial, the sort of thing that happens all the time. When the same teenagers who pried off the plywood come back and set a fire, and one of them is trapped and killed, your report is going to be read by the press, attorneys for the owner and the dead child's family, your municipal attorney, the police and fire departments, and your local prosecutor.

How well you document your actions may have a significant impact on the investigation, potential criminal trial, and civil litigation. The old axiom *"If you didn't write it, you didn't do it"* is true. You must document all of your actions. Good field notes ensure a good final report.

Chapter 5

Writing a Report

Now it's time to expand the information in your notes into complete sentences and paragraphs.

If you're using a department-issued laptop or a paper form provided by your agency, you'll begin by filling in the spaces with information from your notes. Your actual writing process will begin when you come to the space used for your narrative.

If you're writing your report on a blank piece of paper, you'll begin your report with an opening sentence that serves two functions. First, it answers the "5W" questions: **who**, **what**, **when**, **where**, **why**. Second, it establishes your reason to be on the premises. Right of entry clauses in the model codes state that you have the right to be on a premises in order to inspect and enforce the code—not for any other reason. If you can't justify why you became involved in the situation, a judge may eventually dismiss the case. (Sample opening sentences appear on the next page.)

Note that the opening sentence may be different if you're using a laptop to write your report. If you're typing **who**, **what**, **when**, **where**, or **why** information into boxes on the laptop, you may not need to repeat that information in your opening sentence.

The body of your report will appear in a section called the **narrative**. Here you'll describe what you saw and what you did, as well as the statements and actions of victims, witnesses, and suspects.

The narrative will end with a description of outcomes, called the **disposition**. Did you collect evidence, call an ambulance, issue a notice of violation, make a referral to another agency? All of these must be documented.

Actual reports vary, of course. Broadly speaking, they fall into four types, each with its own special features. Understanding these four types saves time and energy because you have a template to work from when you sit down to write. You won't have to worry about what to include or how to organize your information.

The First Sentence

Here's a standard opening sentence you can modify to use in your own reports:

At *time* on *date*, I, *rank, name, ID#, was dispatched to/was conducting an inspection, told about/ at location.*

These examples show how you can adapt the model sentence to your own reports:

At 0815 hours on 4 January 2009, I, Inspector John Brown, was conducting a framing inspection at 301 Crown Place, Smithville.

At 1120 hours on 9 October 2010, I, Inspector Susan Kimura, saw a delivery truck unload drywall and trusses on the street in front of 206 East King Street.

At 1540 hours on 20 August 2010, James Anderson, of 301 Green Street, told me, Inspector Terry Winn, about an illegal boarding house next door, at 303 Green Street.

At 2150 hours on 21 July 2010, I, Inspector Mary Balodis, was dispatched to 1132 Short Street, Junction City, to investigate a deck collapse.

Reminder: If you're filling out a paper form or using a laptop template, you probably won't need to put all this information into your opening sentence (unless your agency or institution asks you to do so).

Names

What about names? Inspectors sometimes wonder what to call a person who couldn't be identified right away. For example, suppose you see a truck, taking water from a hydrant two blocks away. You suspect without a permit. You didn't identify the driver until you caught up with her on a construction site and asked for her driver's license. The general rule is that you can use the person's name right away. In some circumstances you can write, "later identified as Frances Palmore." Note how names are used in these examples:

> At 0910 hours on 16 April 2010, I, Inspector Barry Cameron, saw Billy Pemberton operating the hydrant and taking water from the hydrant at 1520 Camellia Parkway.

> At 0440 hours on November 5 2008, I, Inspector Sally Strathmore, saw a woman (later identified as Karen Huggins) removing paint with a torch from the steeple of the First Baptist Church on Seaman Street.

The Narrative

For many inspectors, telling the story (in technical language, the *narrative*) is the most difficult part of writing a report. Often the story began before you got there. Instead of getting the story in one big chunk, like a TV show, you might get bits and pieces from several people. And they may start by telling you about events that happened in the middle of the story or even near the end.

How do you put all this together efficiently and effectively?

The answer is to use groupings. Remember, you're not writing a Hollywood script. It's perfectly OK (even recommended) to have a separate paragraph for each person's part of the story.

Suppose a landlord has been performing unauthorized work on the boiler without a permit. You might get bits of the story from multiple tenants or the neighbors. Use a separate paragraph for each one.

The examples below show how you could report what the next-door neighbor and the tenant in apartment 2-B told you. (You'll notice that the information is written in bullet style, which you'll learn more about beginning on page 63.)

Mark Grant, who lives next door at 317 Willow Street, told me:

- John West has replaced boilers and other equipment himself, without permits in the past five year

- West is unlicensed

- He performs the work at night on weekends, when City inspectors are not working

- West has paid Grant and other neighbors in cash to help him

Karen Grant, tenant in apartment 2-B and Jason's mother, told me:

- There was no heat in her apartment on four occasions the past month

- John West told her get to it when he got off work that evening

- She would hear him in the boiler room after ten o'clock each time

- West told her he couldn't afford a contractor

And so on, for each person you talk to. You'll see more examples in the next chapter.

Disposition

If you talk to experienced inspectors, you'll probably hear reminders about the importance of the final part of your report, where you tie up the loose ends. Did you issue a notice of violation or order, take a picture, or log evidence? Were there injuries? If so, what did you see, and did an ambulance come? Did you report the incident to other agencies or the police? Were other inspectors or agencies on the scene?

Incomplete details can create problems if a reporter asks for more information or an attorney challenges you in court or at a hearing later on. Ensure that *every report* you write is thorough and complete.

Exercise 4 Write a Report

Instructions: Write a report about the scenario below, following the steps below. When you're finished, compare your report to the sample on page 203.

- Write an opening sentence.
- Write a paragraph describing your arrival at the scene.
- Write a paragraph describing what happened while you were there.
- Write a concluding paragraph.

Scenario:

At 1:30 p.m. today you were dispatched to 11 Clover Lane for overcrowded conditions at the local Hindu Temple social hall. When you arrived, you saw a large number of people entering the building; some were carrying gifts. Blake Chardoury met you at the door and told you one hundred guests were expected at his sister's wedding. He invited you in, pointed to the occupant load placard marked 150 persons, and led you out back, where chairs were set up for the ceremony. Deval Agarwal, a member of the board of trustees, told you John Stem, who lives next door, has complained to him about the lack of parking and of temple guests blocking his driveway. He said they have made a concerted effort to make sure no one parks in front of his driveway, by checking every 15 minutes before and during events. He feels the issue may involve more than parking. You thank both gentlemen and walk next door to speak with Mr. Stem. His driveway is empty, and no one answers the door. You leave a card and return to service, after taking no action.

Exercise 5 Why Is the Report Necessary?

Instructions: No action was needed on your part because no violations were observed. Why is it important to write a report about what happened? Give as many reasons as you can. When you're finished, check your answers on page 204.

Chapter 6

Reports, Notices, Citations, Summonses and Orders

The reports, notices of violation, citations, summonses, and orders you write must meet the minimum legal requirements of your jurisdiction, conform to your department's internal policies, and be written and organized to convey a clear message to the recipient. Each serves a different purpose.

Notices, citations, and summonses are legal documents and contain much of the same information. A *summons* is a legal order for a person to appear before the court. A *citation* is a type of summons that does the same thing. Each state has different requirements based on state statute. *Orders* are statements requiring a person or persons to take action or refrain from something. A *notice of violation* may include an order. Reports capture the details of an incident or situation at a particular time. Let's look the elements of each one.

Reports

Reports document an event, a series of related events or conditions at a particular time. As an inspector, you may be called on to write a report that details your department's inspection and enforcement efforts at a particularly difficult property. Perhaps the owner is claiming his rights have been violated, sanitary conditions are deplorable, necessitating an evacuation order, or unsafe conditions have resulted in injuries or deaths. You may be required to write a report documenting property damage to your department vehicle after an accident. You may be also required to document your actions or those of another person at a meeting or hearing.

Your report should flow in a logical order, typically with an opening statement that establishes the *Why* of the five W's discussed in Chapter Four and includes the other W's. Are you documenting your actions or those of multiple persons? Be consistent in the manner you identify persons. Don't refer to one person as Mr. or Ms. and then identify the others using only their last names. First and last names are best. Indicate rank, position, or title if appropriate.

Think ahead. If the report is documenting an incident that may require follow-up activity, include names, telephone numbers, and addresses to save time. Cell phone numbers are particularly important if persons have been forced out of the property due to fire or other unsafe conditions.

If you are writing an inspection report using a department form, pay particular attention to correctly identifying the legal address. Check the form for completeness. If provided, use the space provided for a narrative description wisely. You may be called upon to testify about the inspection in the future. Your report will help you recall what transpired.

While correct code terminology is important and often a legal requirement, consideration has to be given regarding the technical knowledge of the recipient. The landlord who rents and maintains four apartments in his building probably doesn't know that the correct code term for the leaking fixture in the apartment bathroom is *water closet*. If you use the term in a report or notice of violation he may tell you there isn't a closet in the bathroom. *Toilet* is probably the better term, or use *water closet* and include *toilet* beside it in parentheses. Similarly, construction and electrical codes refer to wires that carry electric current as *conductors*. But fire and property maintenance codes generally use the term *wiring* as in *exposed wiring* or *damaged wiring*. A landlord is likely to associate conductors with trains, so be sure to add *wiring* in parentheses.

Notices of Violation

A notice of violation (NOV) is a written order with the express purpose of gaining compliance with the code. It should be readily apparent to the person who receives it exactly what needs to be done. The model codes promulgated by the International Code Council (ICC), National Fire Protection Association (NFPA), and the International Association of Plumbing and Mechanical Officials (IAPMO) are silent on the minimum required elements of an NOV. However, many states and their political subdivisions have included minimum elements in their amendments. You must comply with the requirements of your adopted code or the NOV may be declared invalid.

Since the purpose of an NOV is to gain compliance, it should include enough information for the property owner or agent to understand what is wrong and what must be done to correct it. In addition to your name, contact information and other standard administrative information, NOV's should include:

- The unsafe or noncompliant condition
- Applicable code section reference
- Statement of what must be done in order to comply
- Date of a follow-up inspection or time for compliance

Example: This would be the minimum information to include, formatted according to your department's usual practices:

Lavatory in second floor men's room leaks

IPMC 504.1

Repair the leaking fixture

Follow-up inspection will be conducted 09/18/2013

Citations or Summonses

Citations and summonses are legal documents that compel appearance before a judicial officer. The power to issue either one, are conferred by the jurisdiction in accordance with state statute and local ordinances. Either is serious business. In some jurisdictions, code violations are criminal misdemeanors. Issuing a summons is legally an arrest, where the accused is released based on a promise to appear in court.

In addition to meeting all of the legal requirements, a clear, concise description of the violation and applicable code section(s) are essential. Judges and attorneys aren't code experts. Boil complex issues down to their essence. Avoid overly complex code terms. Consider including the common term in parentheses beside the code term.

Orders

An order is a written statement from a public official in an official capacity that compels a person to take an action or refrain from one. Orders are usually included in a notice of violation. Orders are sometimes issued prior to a violation occurring, such as when a fire official encounters a brush pile and issues a written order prohibiting open burning. Orders should be concise and should include a reference to the code section authorizing the order if not part of an NOV. Orders should include the applicable code section that is the basis for the order.

Exercise 6 Reports, Notices, Citations, Summonses, and Orders

Instructions: Assume you are an inspector with jurisdiction, who encounters the code violations below. Write an order to address them. When you're finished, check your answers on page 204.

Example:

> Situation: Trap for the lavatory in the second floor men's room leaks.
>
> IPMC Section 504.1 requires plumbing fixtures be maintained free from leaks.
>
> Order: Repair or replace leaking lavatory trap in second floor men's room in accordance with IPMC 504.1.

1. Situation: Clothes dryers in an apartment building are being exhausted into the adjoining trash room.

IPMC Section 403.5 requires clothes dryer exhaust systems shall be exhausted outside the structure, or in accordance with manufacturer's instructions.

> Order:
>
> _____
> _____
>
> _____
> _____

2. Situation: The single required lighting fixture in a common apartment stairway is inoperable.

IPMC Section 402.2 requires stairways to be illuminated at all times.

> Order:
>
> _____
> _____
>
> _____
> _____

3. Thirty-six cardboard boxes of paper records are being stored in the exit stair tower of a bank building. IFC Section 315.2.2

states that combustible materials shall not be stored in exit enclosures.

Order:

_____
_____

_____
_____

4. The manager of an apartment building is maintaining a Dumpster in the marked fire lane behind the building. IFC Section 503.4 prohibits obstruction of fire apparatus access roads.

Order:

_____
_____

_____
_____

5. A homeowner has stopped maintaining the aboveground swimming pool in his back yard. The pool water is green and teeming with insects. IPMC Section 303.1 requires swimming pools to be maintained in a clean and sanitary condition.

Order:

_____
_____

_____
_____

Chapter 7

Objectivity

Opinions have no place in an inspection report. If you're new to report writing, this objectivity requirement may take some getting used to. It's natural to want to state that the owner was irate or the housekeeping in the warehouse was deplorable. And it's tempting to write that the manager is irresponsible and will probably block the exit door the minute that you leave. You may want to say that the manager was disrespectful when you informed him that his storage was too close to the ceiling. In everyday life we often think along these lines.

But inserting these opinions, hunches, guesses, and predictions into an inspection report risks labeling you as unprofessional. Even worse, that kind of writing can make it seem that your actions were personal rather than professional, and it can get you into trouble during an appeal hearing or in court.

A skillful attorney can use vague descriptions ("Conditions in the restaurant were unsanitary") to cast doubt on your judgment, trip you up on the witness stand, or convince the appeals board or judge that you were flying by the seat of your pants during the inspection and couldn't identify code-specific conditions.

Objective (factual) reports make you look professional, and they're especially useful in appeals hearings and in court. After a long time has passed, you may not remember details about what you saw. If they're plainly stated in your report, you'll have no problem testifying. And many inspectors say that good reports can help keep a case from being appealed or landing in court. An attorney who sees that you've convincingly stated the facts may decide not to challenge what you did.

It's best to described conditions in specific terms rather than label them. For example, instead of writing "The lot was overgrown with weeds," which is subjective, you could write "Grass and vegetation exceeded twenty-four inches in height," which describes a clear, measurable code violation.

Here are some examples:

Opinion	Observable Fact
poor housekeeping	combustible waste and refuse is on the floor throughout the house
poorly maintained	the vent pipe is rusted, leaking fuel oil has soaked into the concrete floor, furnace cover is missing
rodent harborage	I observed rodent droppings, damaged fabric, and a nest composed of bits of a paper and fabric in the chair
overgrown	grass and weeds exceeding 24 inches
inoperative vehicle	Chevrolet sedan without tags, current inspection sticker, and wheels, resting on cinderblocks
unsanitary	toilet room walls and floor are soiled, and there is a strong odor of urine and feces

Writing an Objective Report

Judgments regarding lifestyles, living arrangements, cultural customs, or religious practices have no place in an inspection report. However, don't let political correctness hinder you in reporting conditions as they relate to the code in concise, factual terms.

Closely related to objectivity is the issue of *sensitivity*—another hallmark of a professional inspector. Insensitive language should be omitted from your report *unless you're quoting someone's exact words*. Here are some guidelines:

- Use *boy* and *girl* only to refer to children.

- Do not use street slang for minorities or disabled persons.

- Do not use sexually charged language to refer to women (*broad, stacked, bombshell,* etc.)

- Use neutral language when you're referring to people as a group. Appropriate terminology includes "low-income" (rather than "poor" or "lower-class"), "persons with a history of mental illness," "persons with epilepsy," and "gays and lesbians." Street language such as "crazy," "spastic," and similar terminology is not appropriate for an inspector.

Exercise 7 Objectivity

Instructions: Put a check √ in front of the sentences that are objective. Put an X in front of any sentences that lack objectivity. Answers appear on page 205.

____1. Mr. Bo applied for the permit to burn a two-foot square foot house as part of a Chinese funeral ceremony.

____2. The fire was part of a Hindu wedding ceremony.

____3. The house was filthy and reeked of stale spices.

____4. The fire was started by John Wells, who appears to have mental problems.

____5. The grounds are unkempt and unsightly.

____6. From the sidewalk, I observed two inoperable cars in the back yard.

____7. I red-tagged the furnace because it posed a hazard.

____8. The lumber stacks exceeded twenty feet in height and were leaning.

____9. The swimming pool water was green.

____10. I saw twelve cots and four air mattresses in the home.

Chapter 8

What To Omit

Inspectors often worry (and rightly so) about leaving something important out of a report. But it's also true that some things *don't* belong in a report. Here are some examples:

- Opinions (*The contractor thought no one would see the wiring after close-in*)

- Conclusions (*The stack of lumber fell because the forklift operator wasn't paying attention*)

- Generalizations (*The plumbing contractor seems competent*)

- Hunches (*The tenant was probably lying when she filed the complaint*)

- Insensitivity (*Mr. Nagy is obviously crazy and needs to be in an institution*)

You also need to watch for writing practices that don't belong in a modern inspection report. The three practices discussed below can lead to inefficiency and errors.

1. Passive voice

Stick to active voice unless you're describing an action by an unknown person. Passive voice can be wordy and confusing (with the exception noted below).

Here's an example of what can happen if you consistently use passive voice. Suppose you wrote, "Clark was questioned about boarding up the structure Monday morning." Now fast-forward to an appeals board hearing three months later. You may not remember whether it was you or your partner who did the questioning—and your report won't help you.

It's OK to use passive voice when you really don't know who performed an action: "An inoperative Ford sedan was abandoned on the property sometime between March 1 and April 30." Otherwise you should stick to active voice. (You can learn more about passive voice beginning on page 71.)

2. Jargon

Expressions like "The exterior paint is alligatored," "The contractor used a come-along to straighten the wall," and "Seal the opening with safing" can confuse outsiders who read your reports—and they give the impression that you were in too much of a hurry to explain clearly what you did. These sentences are more professional:

> The exterior paint is cracked due to the elements.
>
> The contractor used a winch to straighten the wall.
>
> Seal the opening with an approved fire stop system.

3. Unnecessary repetition

Needless words waste time and leaves you open to factual and grammatical errors. You don't need to write down everything you said when you're questioning a witness or a suspect. Omit expressions like "Then I asked him," "I followed up with," "My next question was."

Compare the two versions below:

> I asked Michelle Wilson how many people were on the deck when it collapsed. She said since she was in the house, she didn't know. I asked if any people had arrived uninvited. She said no. Then I asked her if she could estimate, based on the number of people left in the house. She said there were 10 people left in the kitchen when the fireworks show started. I asked how many people were invited. She

said 20 couples were invited, but some spouses were unable to attend. Then I asked if 20-30 was a good estimate. She said yes. REPETITIOUS

I asked Michelle Wilson how many people were on the deck when it collapsed. She estimated 20–30, based on the number of people who remained in the kitchen when the fireworks show started. BETTER

Sometimes, of course, you'll need to record a person's exact words. It's usually a good idea to record everything that a person involved in a significant incident says, word-for-word, and anything that a witness heard that person say. Use quotation marks in both your notes and your report to signify that you recorded the statements accurately.

Develop the habit of checking your reports over to see how you can improve them. Over time, you'll see a dramatic and very satisfying improvement.

Exercise 8 Rewrite These Sentences

Instructions: Use the guidelines in this chapter to rewrite these excerpts to meet modern report writing standards. You may need to invent some details. Suggested answers appear in the Answer Key on page 206.

1. Robert Webster was ordered by this inspector to extinguish the rubbish fire.

2. James Pate appeared nervous when I asked if he had called for a close-in inspection. I suspected he was lying when he said he thought the general contractor would call it in.

3. This inspector arrived on the job site at 07:15 and found that the sprinkler contractor had not arrived.

4. The homeowner was ordered to remove the bags of trash that were piled on his driveway.

5. John Roberts was observed lighting the controlled burn by this inspector.

6. The electrical contractor probably thought I wouldn't check with the general contractor, when he said the general contractor would call in the inspection.

7. Mr. Wilson became agitated when I informed him that having an inoperative vehicle on his property was a violation of city code. He thought his bluster would frighten me off.

Chapter 9

Quoting Exact Words

Inspection reports usually describe the condition of a structure or premises. Statements from the owner, manager, or tenant generally aren't included in an inspection report unless the inspection is complaint-based. There are times, however, when writing down *who* said *what* will mean the difference between your being perceived the faithful public servant doing your job, and being labeled the aggressor, abusing the power vested in you by the public.

Police officers frequently write down what victims, witnesses, and suspects say. They get a lot of on-the-job training and improve their skills through sheer repetition. But inspectors do this infrequently and may not have opportunities to practice recording statements accurately. So it's important that inspectors acquire the skill through training. Getting the words right is vital. Sometimes it can even mean the difference between an acquittal and a successful prosecution, or prevailing in an appeal. (Getting the punctuation right is just as important because it showcases your professionalism. Instruction on using quotation marks begins on page 125.)

Developing your ability to concentrate is the first step towards learning how to record people's statements accurately. Most people spend most of their time thinking about their own lives and their own problems. In a conversation, they're usually thinking about what they're going to say next. As an inspector, you need to redirect your thinking to the situation at hand, observing and retaining everything that's said.

It's equally important to strengthen your ability to remember. Here's one way to do it: When you watch TV or listen to the radio, try to repeat exactly what you heard. Keep practicing, and strive to increase the number of words you can retain in your memory. After a conversation or a meeting, see if you can repeat what each person said.

Here are three suggestions for accurately recording what you hear when you talk to witnesses, victims, and suspects:

1. Be specific.

"Robert Jones threatened me" isn't good enough. You need to record *exactly* what he said and did: Robert Jones took two steps forward, made a fist, and said, "You'd better watch your back, because I'm not gonna quit until I get you for this." CORRECT

2. Don't shy away from writing down slang, blasphemy, indecent words, and racial slurs when you're quoting a witness or suspect.

Record *exactly* what the person said, word-for-word. Some people use the same phrases and expressions over and over. Your report might provide information that shows the appeals board or court that the person's intentions are contrary to the language in their appeal and their testimony. Inflammatory language and threats need to be included in your report *word for word*.

3. Don't comment or editorialize about what was said.

Observations like "I was shocked" or "I knew she was lying" don't belong in a professional report. Stick to observable facts. Document everything you can. Often a person's actions (trembling, making a fist, asking the same question over and over) will reveal what the person was feeling—and you'll maintain your objectivity.

Exercise 9 What Did They Say?

Instructions: Put a check √ in front of each sentence that effectively records what a witness or suspect said. Mark ineffective sentences with an X. Answers appear on page 209.

____1. Mr. Jones said that the property is being used as a boarding house.

____2. Jon Eastman threatened me with violence if I didn't get off his property.

____3. The couple used offensive language at us when we got out of our city car.

____4. Ms. Jones disapproved of the way the family was living.

____5. The blasting contractor applied for a permit Tuesday September 3, 2013.

____6. Mrs. Jamison said dust from the blasting site has forced her to run through the car wash all the time.

____7. Walt Ramos told me that the rock broke the window Wednesday while he was at work.

____8. Officer Ken Roberts told me the family has been sleeping in their car behind the shopping center.

Chapter 10

How Helpful is OJT?

OJT (on-the-job training) is how professionals in many fields learn their jobs. Talk to successful people in almost any career, and they're likely to say that their higher education was all well and good, but they really learned how to do their jobs by imitating other people at work.

Sometimes that's a good thing, sometimes not. It can mean that professionals are still stuck in the-way-we've-always-done-it instead of updating procedures and policies in light of new research and technology.

Code administration and enforcement is a case in point. Laptops can make report writing much more efficient because inspectors can enter some of the information into boxes instead of writing out whole sentences. But a supervisor who was trained to write reports in pen and ink may not see the benefits of adapting.

The opening sentence in a narrative is one example. In bygone days, when reports were written on blank pieces of paper, it made sense to cram as much information as possible into the first sentence: "At 0842 hours on 8/07/10 I, Inspector Carole Lynch, was dispatched to a deck collapse at 1512 Carmen Boulevard."

But what if your laptop provides spaces for the time, date, type of call, address, and your official ID? There's no need to re-enter them. But the tradition lives on in many agencies.

Four features of good report writing are especially prone to be forgotten by officers who received their training through OJT:

1. Active Voice

There are still people who believe that inspectors instantly become more ethical and objective when they write in passive voice (*The ledger was checked to see what fasteners were used*) instead of active voice (*I checked the ledger to see what fasteners were used*). The truth, unfortunately, is that there are no shortcuts in code administration and enforcement and no easy ways to turn mediocre inspectors into top-notch professionals. (For more about active voice versus passive voice, go to page 71.)

2. Personal Pronouns

The same mistaken belief lingers on about words like "I" and "me": You'll be more objective and professional if you avoid "I" and "me" and write "this inspector" instead. This too is wishful thinking. Think about people you've known who are biased, opinionated, or prejudiced. Could you transform those people just by changing a couple of words in a sentence? The obvious answer is *no*.

3. Bullet Style

Many agencies are discovering that they like bullet style (explained in detail on page 63) because it's easier than writing complete sentences. Other benefits are that bullet style is more compact, easier to organize, and quicker to read—a particular advantage when you're getting ready to testify at an appeal hearing or in court.

But some agencies continue to resist making the change to bullet style. What to do? Stick to the policies your supervisor or agency prefers—but, at the same time, make a resolution to be on the lookout for better ways to write reports. When the time comes for you to be promoted, you'll be ready to show genuine leadership in the area of report writing.

4. Timesaving Word Choices

Report writing is a time-intensive task. Why make the job even more burdensome with unnecessary and time-wasting words? There's no need to write "the abovementioned witness" when you can simply use the name: *Paula Olsen*. And words like *respective* and *individual* can often be omitted:

> The neighbors returned to their respective houses. WORDY
>
> The neighbors returned to their houses. BETTER
>
> Individual residents will receive a street closure notice next week. WORDY
>
> Residents will receive a street closure notice next week. BETTER

To learn more about effective word choices, go to page 179.

Exercise 10 Think about OJT

Instructions: Write a list of helpful things you've learned about code enforcement a) from your own experience and b) from other inspectors. When you're finished, put a check √ in front of each item that you think might be useful to future inspectors.

Chapter 11

Bullet Style

Bullet style has become popular in business writing because it's an efficient and readable way to organize groups of facts. Bullet style is equally useful in inspection reports. Since inspectors tend to use the same headings again and again, bullet style can save time and help eliminate errors. The headings used in bullet style will help you organize your thoughts and remember details you might otherwise overlook.

What does bullet style look like? Actually you've seen it a number of times already in this book, in the reports on pages 20-21. Here's an example:

Anne told me:

- She reported the broken wall switch to the landlord on May 3, 2013.
- Bob Sanchez, the landlord, said he'd have it repaired.
- She smelled smoke and called 911 at 7:00 pm May 10, 2013.
- Firefighters cut power and broke the sheetrock around the switch.
- Mr. Sanchez arrived and threatened to evict her for causing the damage.
- Fire Captain Mike Rizzo called for police.
- Police arrested Sanchez and removed him from the house.

Let's take a look at a paragraph in a zoning inspection report:

I responded to a citizen complaint of an illegal auto body shop, operating from a residence at 4307 Grant Street. From the sidewalk, I observed two sedans with body damage parked in the driveway, and three parked on the street in front of the house. Six steel, fifty-five gallon drums were on the ground behind the detached garage. A large fan installed in the garage window is coated with overspray. Damaged automobile parts are piled beside the garage.

Now let's look at the same information in bullet style:

I responded to a citizen complaint of an illegal auto body shop operating from a house at 4307 Grant Street. From the sidewalk I saw:

- Two sedans with body damage parked in the driveway, and three parked on the street in front of the house
- Six steel, fifty-five gallon drums the ground behind the detached garage
- A large fan installed in the garage window, coated with overspray
- Damaged automobile parts piled beside the garage

Useful Headings for Bullet Style

When I arrived at the site, I saw:

[Name] told me:

When I inspected the building I saw the following violations:

I ordered [Name] to make the following corrections:

I informed [Name] that the following inspections must be completed prior to calling for a final inspection:

Bullet style isn't difficult to learn. You can start by practicing writing down everyday information in bullets. You'll find bullet style extremely useful when you write your reports.

Exercise 11 Using Bullet Style

Instructions: Rewrite the paragraphs below in bullet style. Suggested answers appear on page 207.

1. I arrived at 3316 Fourth Street at 10:15 am. I saw that the residents had evacuated and were assembled at the far edge of the parking lot. Many were in wheelchairs covered with blankets. The fire department was already on the scene pumping water from a hydrant. Leaking water had frozen in the parking lot, causing a severe hazard. I checked in with Chief Murphy at the command post. He asked me to contact Public Works and request a sand truck and the School Board for school buses to temporarily shelter the residents.

2. When I arrived at the building, I saw two water closets, two lavatories, a stainless steel kitchen sink, and a plastic utility tub piled in the yard at the end of the driveway. There was no plumbing permit posted. A white utility van with "Bobby Fix-It" was parked in the driveway.

3. After inspecting Sam's Great American Barbecue, I suspended the permit to operate because of serious violations. The restaurant is without hot water. I observed an infestation of roaches in the kitchen and food storage area. The refrigerator was 44 degrees. Hot food on the buffet table had been maintained at 125 degrees for over four hours.

Chapter 12

Active Voice or Passive Voice?

Earlier in this book, on page 51, you saw that agencies today prefer active voice to passive voice. It might be helpful to take a closer look at both types of sentences to see why agencies are moving away from passive voice.

First, passive voice is old-fashioned; today's professional inspectors tend not use it any more. Second, passive voice can be confusing and inefficient because it doesn't state *who* did *what*. For a better understanding of these problems, let's look at two examples, one in active voice, and the other in passive voice:

> Dickert's service station was inspected and three dispensers were found to be leaking. PASSIVE

> I inspected Dickert's service station and found three dispensers leaking. ACTIVE

An obvious problem with "Dickert's service station was inspected" is that the reader doesn't know *who* inspected it — opening the agency up to problems later on if there's a question about the inspection or the notice of violation or citation.

In the past, criminal justice and code professionals sometimes argued that passive voice ensured objectivity and professionalism. If only it were that easy! Unfortunately, that simply is not true.

In the example you just read, imagine for a moment that the inspector who inspected searched Dickert's service station was lazy and less than honest. He was late for lunch, and rather than wait for the inspection covers to be removed so he could observe the interior of the dispensers, he wrote the report based on minor spills from nozzles by patrons. Will writing in passive voice cause the inspector to perform a competent inspection and generate an honest report in this situation? The obvious answer is *no*.

Be careful, however, not to be fooled into "correcting" sentences that were right in the first place. Make sure a sentence is really passive before you change it. Compare these two examples:

> The unlicensed contractor was questioned. PASSIVE VOICE
> While we were questioning the unlicensed contractor, Officer Brown arrived at the scene. ACTIVE VOICE

"We were questioning" is **active** voice because you know that **we** were doing it. The second sentence does not need to be corrected. (The first one is passive and should be rewritten.)

Finally, note that there are two situations in which passive voice is useful and appropriate. The first is when you don't know who was responsible for a particular act. Take a look at these two examples:

> The plywood was pried off a first floor window of the abandoned building last night. Police found empty beer bottles and cigarette butts on the floor. PASSIVE VOICE acceptable

Since you don't know who broke in and partied there, passive voice is acceptable.

Passive voice can also be useful if you don't want to embarrass a person for something he or she has done:

> Unwashed dishes have been left in the break room three times this week. PASSIVE VOICE acceptable

Exercise 12 Using Active and Passive Voice

Instructions: Put a check √ in front of each active-voice sentence. Mark each passive-voice sentence with an X, and rewrite it in active voice. (You may have to invent names as you rewrite the sentences.) Check your answers against the Answer Key on page 209.

____1. Jones was seen carrying plumbing fixtures into the building.

____2. Jones was carrying a toolbox and a pipe cutter.

____3. Three hydrostatic tests were performed on the automatic sprinkler system.

____4. Patterson was looking in his wallet for his contractor's license.

____5. Both witnesses were questioned.

____6. Finch hesitated and looked at his wife when I asked for his license.

____7. Chief Clancy and Major Hansen rewrote the procedure.

____8. The policy was rewritten two years ago.

____9. I was hoping to take a week of vacation in late August.

____10. The wallet was found under the driver's seat.

____11. The mayor will be attending Lieutenant Cohen's retirement ceremony.

____12. Luis is interested in plan review.

____13. Scientists in materials labs are being paid top salaries right now.

____14. Three years ago, Luis was working in a low-paying service job.

____15. He was told there wasn't much of a future for him there.

Chapter 13

Online Resources for Code Inspectors

The Internet provides many free websites about writing that are useful to code inspectors. The websites listed here will prove useful throughout your career, so it's a good idea to bookmark and use them often. All are free.

1. www.Dictionary.com

Besides offering definitions, this website compares what various dictionaries say about a particular word or phrase, and it sometimes offers usage notes.

2. www.PlainLanguage.gov

Jargon and gobbledygook waste time, create confusion, and make a bad impression on your readers. This government-sponsored website provides many easy-to-use resources to help you write more clearly and efficiently.

3. www.YourPoliceWrite.com

Although the site was originally created for criminal justice officers, it also provides a thorough review of grammar, usage, and special issues related to code enforcement report writing.

4. www.WritewithJean.com

This website provides ongoing instruction about a wide variety of writing issues.

5. owl.English.Purdue.edu/owl

(Note that there's no "www.") This is Purdue University's on line writing lab. It contains information on technical writing, including engineering reports. It also has links to videos on writing.

6. Workspace.Office.Live.com

(Note that there's no "www.") This is a career-building website that allows you to collaborate with colleagues. If you're working on a job-related writing project, you can post your draft here and allow colleagues to log in to make additions and edits. There's no need to email drafts to one another: Everything is securely (and privately) stored online. This is a great professional tool that you'll use often as you move up the career ladder.

7. www.FlipDrive.com

Here's another career-building website. You can securely store documents and photographs here. If you move to another computer, just log on to access a project you're working on. You don't need to carry a flash drive around—and you don't have to worry about losing it.

8. www.Evernote.com

Say good-bye to Post-It notes—although they're handy, they're also easily lost. This free, privacy-protected website sorts and stores any information you want to save. You can access the information from any computer with Internet access. Evernote allows you to clean out your desk and set up a quick, reliable system to find important information.

9. www.Passpack.com

This isn't really a writers' website, but it's a lifesaver for many professionals. You can securely store passwords here, free of charge, and access them from any computer with Internet access. This is a great website if you have accounts with many websites, and it's especially useful if you travel often: You don't have to worry about carrying (and possibly losing) a list of passwords.

10. www.InspectorWriteRight.blogspotcom

You'll find writing tips and other topics of interest for code inspection professionals.

Exercise 13   Exploring Online Resources

Instructions: Choose two of these websites and explore what they have to offer you. Then write a brief explanation of how you would use each website you selected.

Section II

Solving Sentence Problems

Chapter 14

Fragments

The first requirement for a sentence also happens to be the most important requirement: Completeness. English grammar is full of obscure rules that you might be able to break without getting caught. But writing an incomplete sentence (also called a fragment) is a serious error that most good readers will notice immediately—to your detriment.

Fortunately there's an easy way to ensure that your sentences will probably meet this **completeness** requirement: Start every sentence with a person, place, or thing. To put it differently: If you're unsure of your writing skills, take the safe route. Avoid writing complicated sentences that can get you into trouble.

Take a look at the two paragraphs below. They state the same information, but the first version contains several fragments. The second version is less fancy—but every sentence is correct.

Version 1 (fragments are underlined):

We separated Jennings and Cooper. I took Jennings into the kitchen. <u>Asked her what had happened.</u> <u>Trying to decide who was telling the truth.</u> Jennings told me she replaced the handrail the day Cooper reported it missing. She thinks Cooper is trying to break the lease. Removed the handrail himself. She suspected that trouble was coming. <u>Because he didn't seem like himself lately.</u> <u>Seemed like something was eating at him.</u>

Version 2 (all sentences are correct):

We separated Jennings and Cooper. I took Jennings into the kitchen. I asked her what had happened. I was trying to decide who was telling the truth. Jennings told me she replaced the handrail the day Cooper reported it missing. She thinks Cooper is trying to break the lease and removed the handrail himself. She suspected that trouble was coming because he didn't seem like himself lately. It seemed like something was eating at him.

A Closer Look at Fragments

The following tips will help you avoid writing fragments (incomplete sentences):

- Remember that most fragments appear at the start of a paragraph or the beginning of a section of a report. Double-check those spots for completeness.
- Be careful with sentences beginning with words like *first*, *next*, and *finally*: Fragments often creep in there.
- Check every sentence that begins with an *–ing* word: Such sentences are notorious for turning into fragments.
- Starting each sentence with a person, place, or thing is good insurance against fragments.
- Avoiding "red flag" words at the beginning of a sentence is also good insurance. In general, avoid starting sentences with *like*, *who*, *which*, and *such as*.

What Do Fragments Look Like?

Here is a report containing several fragments (underlined):

Palm Court is an assisted-living facility administered by the Methodist Church. Administrators at Palm Court met with representatives from several community agencies to discuss enhancing security at the facility. They noted several concerns.

<u>First, unauthorized visitors.</u> Palm Court encourages visits from family and friends. Because the staff wants to maintain a welcoming atmosphere, there is no sign-in procedure for visitors. Several thefts

have occurred because outsiders come and go freely. <u>Entering and leaving the building without being stopped and questioned.</u>

<u>Second, inadequate locks.</u> In the past, Palm Court did not install sturdy locks on apartment doors. <u>Worrying that residents would accidentally lock themselves out.</u> As a result, residents are not adequately protected against intruders.

<u>Third, untrained staff.</u> Palm Court has done a good job recruiting caring workers who are sensitive to the needs of elderly residents. But staff members have not been taught how to maintain a secure facility. <u>Having proven procedures to protect residents and their possessions from harm.</u>

And here is the same report with the fragments corrected (in *italics*):

Palm Court is an assisted-living facility administered by the Methodist Church. Administrators at Palm Court met with representatives from several community agencies to discuss enhancing security at the facility. They noted several concerns.

The first problem is unauthorized visitors. Palm Court encourages visits from family and friends. Because the staff wants to maintain a welcoming atmosphere, there is no sign-in procedure for visitors. Several thefts have occurred because outsiders come and go freely, entering and leaving the building without being stopped and questioned.

The second problem is inadequate locks. In the past, Palm Court did not install sturdy locks on apartment doors. *Administrators worried that residents would accidentally lock themselves out.* As a result, residents are not adequately protected against intruders.

The third problem is untrained staff. Palm Court has done a good job recruiting caring workers who are sensitive to the needs of elderly residents. But staff members have not been taught how to maintain a secure facility. *It's important to establish proven procedures to protect residents and their possessions from harm.*

Exercise 14 Fragments

Instructions: Put an X in front of each fragment, and then rewrite it so that it is complete. When you're finished, check your answers on page 210.

_____1. Williams Plumbing needs a final inspection for 9:30 this morning.

_____2. Needs to backfill the trench before tonight's rain.

_____3. Although, the work has been complete for three days.

_____4. Noticing that he has a pattern of requesting Friday inspections at the last minute.

_____5. Other contactors have complained.

_____6. They say it's unfair that Williams doesn't have to follow the rules.

_____7. Williams uses impending weather or Monday holidays as an excuse.

_____8. Trying to manipulate the system.

_____9. Which creates disorder in the scheduling process.

_____10. Inspector Link and I talked with Williams about scheduling inspections twenty-four hours in advance.

Chapter 15

Run-on Sentences

Of all the mistakes a writer can make, run-on sentences are among the most serious. So what's a run-on sentence, and how can you avoid making this error?

A "run-on" is two sentences joined together without a period:

> The dog barked in the middle of the night, Wilson looked out the window. RUN-ON

> The dog barked in the middle of the night. Wilson looked out the window. CORRECT

> A fight broke out near the food truck, two electrician's helpers were arguing about a chicken sandwich. RUN-ON

> A fight broke out near the food truck. Two electrician's helpers were arguing about a chicken sandwich. CORRECT

The problem with a run-on is that it doesn't stop when it's supposed to. (Think of a car engine that "runs on": It's the same problem—not stopping when it's supposed to.)

Some officers wrongly think that any long sentence is a "run-on." Not true! Long sentences are perfectly correct *as long as* there's a period in the right place. Here's a long sentence from Thomas Jefferson's first Inaugural Address. It's grammatically correct and doesn't need any corrections:

> And let us reflect that, having banished from our land that religious intolerance under which mankind so long bled and suffered, we have yet gained little if we countenance a political intolerance as despotic,

as wicked, and capable of as bitter and bloody persecutions. CORRECT

Actually some run-ons are quite short. They're still wrong if the period is missing. (Note, by the way, that some teachers and editors use the term "fused sentence": It's the same thing.)

Jane was frightened, she hid in the closet. RUN-ON

Jane was frightened. She hid in the closet. CORRECT

I pushed, the door opened. RUN-ON

I pushed. The door opened. CORRECT

Avoiding Run-on Sentences

Here are a few suggestions to help you avoid run-ons:

- Remember that "it" often starts a new sentence:

I searched his locker, it was empty. RUN-ON

I searched his locker. It was empty. CORRECT

- Practice distinguishing between "extra ideas" (which end in commas) and sentences (which require periods or semicolons):

When the alarm went off, we ran to the back door. CORRECT ("When the alarm went off" is an extra idea)

The alarm went off, we ran to the back door. RUN-ON
The alarm went off. We ran to the back door. CORRECT

- Study pages 97-102, which cover semicolons and periods, to gain confidence with punctuation.

Exercise 15 Identifying and Correcting Run-on Sentences

Instructions: Insert periods or semicolons where they're needed. Some sentences don't need corrections. When you're finished, check your answers on page 210.

1. Knudsen saw someone going through the construction materials, no charges were filed.

2. When I entered the sun porch, I saw the condemned sign on the door.

3. The permit counter was crowded Duran walked out.

4. Culpepper said the blaster had a snake tattoo, gold hoop earrings, and two missing front teeth.

5. Carr insisted that because he was the owner, he could build a rear deck in any manner that he chose.

6. I approached the dog, it growled at me.

7. Nieminen said she heard screeching brakes and a thud, she told her husband to go outside to look.

8. One car had a dented fender, the other was undamaged.

9. No one enjoys working holidays, however in our profession it's often necessary.

10. I talked to the lieutenant, then I went straight to the gym.

Chapter 16

Misplaced Modifiers

The term "misplaced modifier" may sound like English teachers' jargon, but it points to a real-world writing problem you should avoid in your reports. (Another name for this problem is "dangling modifier.")

"Misplaced" means *hanging*, and a "modifier" is a *description*. So a "misplaced modifier" is a description in the wrong place. Most misplaced modifiers are easy to spot because they sound ridiculous. Take a look at these examples:

> Throughout the house, Inspector Jones photographed the illegal electrical work. MISPLACED MODIFIER
>
> I spotted the undersized lag bolts searching for fasteners. MISPLACED MODIFIER
>
> I saw a disconnected smoke alarm walking through the bedroom. MISPLACED MODIFIER

Here are the corrected sentences:

> Jones photographed the illegal electrical work throughout the house. CORRECT
>
> While searching for fasteners, I spotted the undersized lag bolts. CORRECT
>
> Walking through the bedroom, I saw a disconnected smoke alarm. CORRECT

Sometimes misplaced modifiers are harder to spot. To most people, this sentence probably looks correct on first reading—but it isn't:

> Questioning the tenant, he said his landlord told him it was okay to store the equipment in the attic. MISPLACED MODIFIER

There are two problems with the sentence. First, the tenant didn't do the questioning. Second, the sentence doesn't specify who did.

The omission might create a problem in an appeal hearing or in court, when it's important to identify all the parties involved.

Here's the corrected sentence:

> When I questioned the tenant, he said his landlord said it was okay to store the equipment in the attic. CORRECT

Be careful when you start a sentence with an *-ing* word: Often it will contain a misplaced modifier. If you do start a sentence with an *-ing* word, check to make sure it's clear who did what.

Exercise 16 Misplaced Modifiers

Instructions: Make any corrections that are needed in these sentences. Not all sentences need corrections. When you're finished, go to page 211 to check your answers.

1. Holding the pipe wrench unsteadily in his right hand, it dropped off the scaffolding and struck the superintendent in his shoulder.

2. We spotted the blaster's truck driving down Parker Avenue.

3. After questioning Li, I left my card and asked him to call me if he recalled anything else about the deck collapse.

4. Scattered around the room we saw parts of the circuit breaker panel that had shorted out.

5. Inspector Pierarski found the little girl hiding behind a rosebush in the back yard.

Chapter 17

Parallelism

Parallelism can become an issue when you write a sentence about three or more things or events. More often than not, the sentence will take a wrong turn when you reach the last part. To see how parallelism works, compare these two versions of the same sentence:

The room had mold on the ceiling, walls, and the floor was buckling.
INCORRECT
The room had mold on the ceiling and walls. The floor was buckling.

CORRECT

Most parallelism problems are easy to fix. Usually you can break a larger sentence into two shorter ones.

Spotting Parallelism Problems

First you need to know how to spot parallelism problems. Here are some suggestions:

- Pay special attention to sentences with three parts
- Remember that the third part is usually the problem
- Try thinking of the sentence as a little poem

Let's try that "little poem" strategy with the previous example so you can see how it works:

The room had mold on

the ceiling

the walls

the floor was buckling

You can quickly see that "the floor was buckling" doesn't match the other two parts. The easiest solution is to make two sentences, as noted before: *The room had mold on the ceiling and walls. The floor was buckling.*

The sentence about the room would work if there were three items that matched "had mold on":

The room had mold on

the ceiling

the walls

the floor

Here's how the corrected sentence would read: *The room had mold on the ceiling, walls, and the floor.*

Let's try another example:

Inspector Hines and I reviewed the plans, inspected, and approved the new shopping center.

To check this sentence for correctness, think of it as a little poem:

Inspector Hines and I

reviewed the plans

inspected

approved

You can see that all parts match: reviewed the plans, inspected, approved. The sentence is correct.

Being careful with parallelism gives you writing a more professional look. It's an important skill for officers to master.

Exercise 17 Parallelism

Instructions: Make any corrections needed in the sentences below. Not all sentences need corrections. Check your answers on page 212.

1. Connors told me she locked the door, turned on the alarm, and a neighbor had the alarm code.

2. Ricky Lopez finished the basement without a permit, placed a construction Dumpster on the street without permission, and said the mayor wants him to bid on city contracts.

3. Each applicant must submit a birth certificate, take a physical examination, and they must come in for an interview.

4. In recent years we've been recruiting more women, minorities, and taking a harder line on racism and sexism.

5. Always check your reports for accuracy, correct spelling, and completeness before you submit them.

Section III

Professional Sentences

Chapter 18

Writing Effective Sentences

As an inspector, you'll need to write clear, error-free sentences that sound professional. When you're looking ahead to promotion, effective sentences are even more important.

But mastering grammar can be a daunting and time-consuming task. What busy inspector has time to grapple with adverbial clauses, correlative conjunctions, and appositives? Suppose, though, that you had to master only four sentence patterns?

Professional Sentence Patterns		
Type of Pattern	**Special Words**	**Typical Sentence**
subordinate conjunction (Comma Rule 1)	*if, when, because, although*, and similar words	Because of the economy, fewer building permits were processed.
coordinate conjunction (Comma Rule 2)	FANBOYS words: *for, and, nor, but, or, yet, so*	I suggested hiring a consulting engineer, but Caffrey refused to spend the money.
interrupter (Comma Rule 3)	*who, which*	Inspector Jagger, who joined the department last year, will head the new project.
semicolon	none required	The code law goes into effect today; it will enhance fire safety.

A Closer Look

As the chart on page 95 demonstrates, three of the sentence patterns correspond to the three Comma Rules you already learned on pages 103-16. Semicolon sentences are even easier to learn. Semicolons are like periods: The only difference is that you lower-case the first word after a semicolon.

In the following pages you'll learn the special words and punctuation needed for each type of sentence, and you'll have a chance to practice applying what you've learned.

Chapter 19

Periods and Semicolons

Professional Sentence Patterns		
Type of Pattern	Special Words	Typical Sentence
subordinate conjunction

(Comma Rule 1) | *if, when, because, although,* and similar words | Because of the economy, fewer building permits were processed. |
| coordinate conjunction

(Comma Rule 2) | FANBOYS words: *for, and, nor, but, or, yet, so* | I suggested hiring a consulting engineer, but Caffrey refused to spend the money. |
| interrupter

(Comma Rule 3) | *who, which* | Inspector Jagger, who joined the department last year, will head the new project. |
| **semicolon** | **none required** | **The new code goes into effect today; it will enhance fire safety.** |

For most writers, periods are the easiest punctuation marks. You already know that sentences end with periods, and that most abbreviations are followed by periods (although some organizations no longer use them).

When you're uncertain about an abbreviation, you can check their website or look at their stationery to see whether periods are needed. UNICEF, IBM, and NASA are examples of organizations that do not use periods. Mr., Dr., Sgt., and similar titles do use periods. If you're uncertain, check the dictionary or visit www.Dictionary.com.

Here's one more piece of information about periods that may prove useful: Space once (not twice) after a period when you're typing. Today's computers are sophisticated typography machines, and the old rules about typewriters no longer apply.

Semicolons

Once you know how to use a period at the end of a sentence, you also know how to use a semicolon: Just change a period to a semicolon, and lower-case the next letter (unless it's a name with a capital letter). Please note that what you *don't* do is pick out a long sentence, find the midpoint, and stick a semicolon there.

Note these examples:

Garrett unlocked the door. We entered the house. PERIOD
Garrett unlocked the door; we entered the house. SEMICOLON

Mark tried hiding the car keys. Judy found them anyway and took his car. PERIOD
Mark tried hiding the car keys; Judy found them anyway and took his car. SEMICOLON

Semicolons are easy to use, and they give your reports a professional look that impresses readers. Forget anything you've heard about elaborate rules for using semicolons. All you need to do is find two sentences that seem to go together.

Take a look at these examples:

Patterson walked the job site. Building materials blocked his path. CORRECT
Patterson walked the job site; building materials blocked his path. CORRECT

I arrived for the inspection at seven o'clock. The contractor was already gone. CORRECT

I arrived for the inspection at seven o'clock; the contractor was already gone. CORRECT

Using Semicolons Effectively

Here are a few suggestions for using semicolons effectively:

- Use semicolons sparingly: One semicolon per paragraph, or one per page in a short writing task.

- Never use semicolons to divide sentences. A semicolon is like a period, not a comma.

- Think of a semicolon as a way to join two sentences into one big one with one capital letter.

- Remember that semicolons are just like periods. You never have to use a semicolon between two sentences. A period will always work.

- Don't be intimidated by semicolons. Any time you have two sentences that are related in some way (and that's most of the time!) you can change a period to a semicolon.

Semicolon or Comma?

Don't try joining sentences with a comma unless you're using Comma Rule 2 (see page 109): Use a semicolon instead (or a period with a capital letter). Here are a few helpful tips:

1. *It* often starts a new sentence and needs a semicolon (or a period and a capital letter).

I like my new laptop, it makes writing easier. INCORRECT

I like my new laptop; it makes writing easier. CORRECT

I like my new laptop. It makes writing easier. CORRECT

2. *However, then, therefore,* and similar words can't be used with a comma to join sentences. Again, use a semicolon (or a period and a capital letter). Only seven words in the English language can be used with a comma to join sentences: *For, and, nor, but, or, yet, so.* (Go to page 119 to learn more about Comma Rule 2 and these FANBOYS words.)

Lister climbed down from the scaffold, then he showed me his permit. INCORRECT

Lister climbed down from the scaffold; then he showed me his permit. CORRECT

Lister climbed down from the scaffold. Then he showed me his permit. CORRECT

Exercise 18 Using Semicolons

Instructions: Read the paragraphs below. In each paragraph, choose two sentences to combine with a semicolon. Check your semicolon sentences against the Answer Key on page 212.

Luther Shalit is a math tutor in the community college building trades program. He helps students learn elementary algebra and geometry. I've seen positive changes since he became a tutor. Luther is proud of his knowledge and happy to be doing something useful. Luther has always been interested in mathematics.

Captain Gephardt asked Linda Hammond to talk to us. She described her work as a Resource Officer at Penny Lane Middle School. She feels she's making a positive difference there. Discipline at the school has improved since she was assigned there. Students trust her and come to her for advice. She discusses substance abuse, family problems, and conflict resolution with students and faculty.

Exercise 19 More Practice with Semicolons

Instructions: Read the sentences below and insert semicolons (or periods and capital letters) where necessary. Check your sentences against the Answer Key on page 213.

1. Vacant buildings are a big problem, the number has increased with the poor economy.

2. The office was quiet during the weekend although a few contractors called for inspections of emergency work.

3. I'm going to interview Davis this weekend, he may have some information about the building collapse.

4. I looked for the camera, however it wasn't there.

5. We're looking for issues that might come up in the accreditation review, such as expired certifications and improper recordkeeping.

6. Our agency is planning a series of events to familiarize the community with our personnel and services.

7. The evaluation was a pleasant surprise, we received an excellent rating in several categories.

8. I found an open receptacle in the living room, Shipton found a cross-connection in the bathroom.

9. After the owner blocked the exit the second time, I issued him a summons.

10. The house is equipped with a silent intrusion alarm, furthermore there are bars on the windows and doors.

Chapter 20

Commas

Three basic rules will cover most of the commas you will use in your reports.

Rule 1

Use a comma whenever a sentence begins with an extra idea.

> **Because the meter had been pulled,** I called for a light unit from the fire department.

If the extra idea is at the back, omit the comma:

> I called for a light unit because the meter had been pulled.

If the extra idea at the beginning is very short, you may omit the comma:

> **Last night** we closed a dozen open permits.

Rule 2

Use a comma when two sentences are joined by *and* or *but*.

> I called Jennings, and I wrote him a letter.

> Wilson remembered the inspection, but she didn't recall the date.

If you don't have two complete sentences, omit the comma:

> **I called Jennings** and wrote him a letter.

> Wilson remembered the inspection but not the date.

Comma Rule 2 can also be used with five additional words: *for, nor, or, yet, so*. **For** in this context means "because": *Inspector Danson suddenly turned around, for she sensed that someone was approaching.* Most of the time, however, you need to focus only on *and* and *but*.

Rule 3

Use a comma in front and another one in back when a sentence contains an interrupter (a word group that disrupts the sentence flow).

> Inspector Bolton, **who was my role model**, is retiring next week.
>
> Park Street, **which is closed to traffic today**, will reopen tomorrow.
>
> Our new headquarters, **scheduled to open in March**, will be much more comfortable.

Omit the commas for essential information:

> Inspectors who earned certifications will be honored in the ceremony.

Tip: Reading the sentence aloud is a big help with Comma Rule 3 commas. You'll hear your voice drop and then go up again. Try it!

Chapter 21

Comma Rule 1

Professional Sentence Patterns		
Type of Pattern	Special Words	Typical Sentence
subordinate conjunction (Comma Rule 1)	*if, when, because, although*, and similar words	Because of the economy, fewer building permits were processed.
coordinate conjunction (Comma Rule 2)	FANBOYS words: *for, and, nor, but, or, yet, so*	I suggested hiring a consulting engineer,, but Caffrey refused to spend the money.
interrupter (Comma Rule 3)	*who, which*	Inspector Jagger, who joined the department last year, will head the new project.
semicolon	none required	The new code goes into effect today; it will enhance fire safety

Comma Rule 1 covers subordinate clauses (word groups beginning with *subordinate conjunctions* such as **if, when, because, although**) as well as prepositional phrases (word groups beginning with *prepositions* such as **of, in, by, for, with, to**). (Formally these are called "complex sentences.")

Using Comma Rule 1 Effectively

Here are a few suggestions for using Comma Rule 1 effectively:

105

- Never place a comma after a subordinate conjunction.

The street was closed because, we were getting ready for the parade. INCORRECT

The street was closed because we were getting ready for the parade. CORRECT

- Don't confuse extra ideas with sentences.

I'd like to go back to college. **Although, this might not be a good time**. INCORRECT

I'd like to go back to college **although this might not be a good time**. CORRECT

- Use a comma if the extra idea is at the front of the sentence (*not* the back).

I saw the driver toss something out the window, **when he spotted me**. INCORRECT

I saw the driver toss something out the window **when he spotted me**. CORRECT

When he spotted me, I saw the driver toss something out the window. CORRECT

Exercise 20 Comma Rule 1

Instructions: Use Comma Rule 1 to insert commas where needed. Not every sentence needs commas. When you're finished, check your answers on page 214.

1. While Inspector Josephs called for an ambulance I disconnected the nail gun from the compressor.

2. The fight began when Todman insulted Jeffers.

3. Inspector Peters impressed the jury although he was nervous about testifying.

4. Because no shrubbery was broken I knew the tenants didn't jump from an upstairs window.

5. If you talk to Wilson in the bedroom I will interview his wife in the kitchen.

6. He has been in business since January 2004.

7. Because I suspected he wasn't licensed I called the clerk's office.

8. We routinely check for business licenses after the permit has been issued.

9. When the fight broke out in apartment Inspector Cary radioed for the police.

10. Although the surveillance camera wasn't working the police found two eyewitnesses who saw the incident.

Chapter 22

Comma Rule 2

Professional Sentence Patterns		
Type of Pattern	Special Words	Typical Sentence
subordinate conjunction (Comma Rule 1)	*if, when, because, although,* and similar words	Because of the economy, fewer building permits were processed.
coordinate conjunction **(Comma Rule 2)**	**FANBOYS words:** *for, and, nor, but, or, yet, so*	I suggested a consulting engineer, but Caffrey refused to spend the money.
relative pronoun (Comma Rule 3)	*who, which*	Inspector Jagger, who joined the department last year, will head the new project.
semicolon	none required	The new code goes into effect today; it will enhance fire safety.

Comma Rule 2 covers sentences combined with the words **and** or **but**. (Formally these are called "compound sentences.")

Actually there are seven words for Comma Rule 2: **and/but** are by far the most common. The word FANBOYS is a good memory device to remember all seven Comma Rule 2 words: **For And Nor But Or Yet So.**

"For" has a special meaning in a Comma Rule 2 sentence: It's much like *because*.

> We've been expecting you, for Colonel Mays said you'd be visiting us soon. CORRECT

Using Comma Rule 2 Effectively

Here are a few suggestions for using Comma Rule 2:

- Never use a comma directly after a coordinate conjunction.

> I was nervous at first but, I soon got over my fears. INCORRECT

> I was nervous at first, but I soon got over my fears. CORRECT

- The seven coordinate conjunctions (FANBOYS words) are the only words you can use with a comma to join two sentences. Use periods or semicolons with other words.

> Bill examined the breaker panel, **then** he looked at the receptacles. INCORRECT

> Bill examined the breaker panel. **Then** he looked at the receptacles. CORRECT

> The promotion I'm interested in pays well, **therefore,** I expect a lot of competition. INCORRECT

> The promotion I'm interested in pays well. **Therefore,** I expect a lot of competition. CORRECT

- *And/But* are the most common Comma Rule 2 words. You'll rarely need to think about the other five FANBOYS words.

Exercise 21 Comma Rule 2

Instructions: The key to Comma Rule 2 is thinking about the words *and* and *but*. Check to see if there's a sentence before and after *and/but*. If that's the case, insert a comma. (Not every sentence needs a comma.) When you're finished, check your answers on page 215.

1. Bill Nelson stamped the plan and called the architect with the bad news.

2. I talked to Jerry Whitman and Inspector Barthes questioned his wife.

3. Cashin produced a key but couldn't open the door.

4. The policy makes sense but we can't implement it this year.

5. The treads and riser aren't uniform and the handrail is too low.

6. The underground main failed both hydrostatic tests and the thrust blocks are missing.

7. The shelter is overcrowded and does not provide enough services for displaced families.

8. I got out of my car and called for a police officer.

9. We questioned the subcontractor but no one heard anything unusual that night.

10. I looked for rebar but didn't see any.

Chapter 23

Comma Rule 3

Professional Sentence Patterns		
Type of Pattern	Special Words	Typical Sentence
subordinate conjunction (Comma Rule 1)	*if*, *when*, *because*, *although*, and similar words	Because of the economy, fewer building permits were processed.
coordinate conjunction (Comma Rule 2)	FANBOYS words: *for, and, nor, but, or, yet, so*	I suggested hiring a consulting engineer, but Caffrey refused to spend the money.
interrupter (Comma Rule 3)	*who, which*	**Inspector Jagger, who joined the department last year, will head the new project.**
semicolon	none required	The new code goes into effect today; it will enhance fire safety.

Use Comma Rule 3 when a word or group of words interrupts a sentence. In most cases you'll use two commas, and changes in your voice will tell you where the commas go:

Your mission, Jim, is to investigate the building collapse. CORRECT

Your next assignment, which you'll find challenging, is to chair the accreditation committee. CORRECT

Both contractors, Watson and Turner, are losing their licenses tomorrow. CORRECT

I spoke to Lily Roberts, from the permit counter, who told me she'd mentioned her suspicions about him to her supervisor several times. CORRECT

Using Comma Rule 3 Effectively

Here are a few suggestions for using Comma Rule 3 effectively:

- Listen to your voice. Use the commas when your voice changes.
- Use two commas, not one, in most sentences.
- Remember that often (but not always), Comma Rule 3 sentences include a *who* or *which* clause.

Mrs. Jones, who told me she'd called 911, said she smelled smoke at approximately 10:30. CORRECT

Irene Withers, who works in the planning department, said she saw the runner leave the site plan at the counter. CORRECT

Exercise 22 Comma Rule 3

Instructions: Read each sentence aloud, listening for a voice change. Insert commas where needed. Check your answers against the Answer Key on page 215.

1. The plan tracking software that we purchased last month is already out of date.

2. Inspector Rice who teaches in the academy part-time has some good suggestions about preparing for the state certification exam.

3. During the winter when many homeless people migrate to Florida the crime rate increases here.

4. Our new permit counter which opened last month is better organized and more secure.

5. Abigail Gleason who is the designer of record visited the job site yesterday.

Exercise 23 Practice with Comma Rules 1, 2, and 3

Instructions: Use all three rules to place commas in these sentences. Not every sentence needs commas. When you're finished, check your answers on page 216.

1. As I approached the house I heard a nail gun.

2. Linda grabbed her son's hand and they ran down the street.

3. Linda grabbed her son's hand and ran down the street.

4. Paul who just graduated from the code academy is planning to go back for a degree.

5. I went back to Porter Street because I had more questions for Mrs. Smith.

6. Bailey's uniform which should have been soiled was suspiciously clean.

7. Menzies arrived at the meeting on time although traffic downtown was moving slowly.

8. The permit counter closed for repairs this week, it will reopen on Tuesday.

9. Glenn was afraid of confrontation at inspections at first but soon overcame his fears.

10. He spent extra time in the codebook and asked Inspector Kelly to work with him.

Section IV

Mastering English Usage

Chapter 24

Apostrophes

Inspectors need to know two ways to use apostrophes: in **contractions** (*can't, didn't, won't*), and in **"of" ideas**: *Mary's uniform* (uniform of Mary), *an architect's office* (office of the architect), *Wednesday's meeting* (meeting of Wednesday).

Apostrophes DO NOT mean "more than one." Note these examples:

> Neighbors called the police when they heard gunshots. NO APOSTROPHE

> The Browns live on the next block. NO APOSTROPHE

> The Browns' house has a swimming pool. APOSTROPHE: house of the Browns

> Karen's codebook is on her desk. APOSTROPHE: book of Karen

> There are two Karens in my code academy. NO APOSTROPHE

There's one exception to the "no plurals" rule: Apostrophes are used in the plurals of numerals and letters: *10's and 20's, p's and q's*. You'll learn more about these apostrophes later.

Where Does the Apostrophe Go?

Before the *s* or after the *s*? It depends on how the word is spelled. Apostrophes always go after the **last letter** of a word or name. If you know how to spell the word or name, you know where the apostrophe goes:

> John   **John's** injuries aren't serious.

Louis   **Louis'** car was stolen.

Mr. Brown   Mr. **Brown's** story needs to be checked.

The Browns   The **Browns'** neighbors called 911.

baby   The **baby's** mother disappeared.

babies   We're collecting **babies'** clothing for the charity drive.

family   Inspector Clay is inspecting the **family's** new house.

families   Both **families'** houses were damaged.

woman   I heard a **woman's** voice on the phone, but I couldn't identify it.

women   **Women's** roles in code enforcement have expanded over the years.

boy   A **boy's** bicycle was found in some shrubbery.

boys   The **boys'** teacher is on paid leave.

Sometimes apostrophes are needed in time expressions:

a **day's** pay   (pay of a day)

two **days'** absence   (absence of two days)

a good **night's** sleep   (sleep of a good night)

three **years'** experience   (experience of three years)

a **week's** vacation   (vacation of a week)

If you don't have an "of" expression or a contraction, don't use an apostrophe:

The **Johnsons** sent me a birthday card.

The **Johnsons'** birthday card surprised me.

My **family's** vacation wasn't long enough.

Having fun together keeps **families** strong.

Contractions

Apostrophes represent omitted letters in contractions: *don't, can't, won't*. Be careful with spelling. For example, in *don't* the apostrophe replaces the missing "o" in *not*.

I am getting ready for my trip to Cleveland.

I'm getting ready for my trip to Cleveland.

Joe is going with me.

Joe's going with me.

Possessive pronouns (like *his*) don't get apostrophes:

That book is **hers**, and this one is **mine**.

Florida is seeing a decline in **its** population.

The Acme Corporation doubled **its** profits last year.

Is that beautiful car **yours**?

It's has only one meaning, a contraction of *it is:*

I won't need a ride home unless **it's** raining.

It's difficult to find a suitable gift for my mother-in-law.

When *its* is possessive (like *his*), omit the apostrophe:

My favorite shirt is missing two of **its** buttons.

Our town got more than **its** share of rain last week.

There's one more way to use apostrophes. When you're writing the plural of a numeral or a letter, use an apostrophe:

Dot your i's and cross your t's.

The cashier gave me my change in 1's and 5's.

During the 60's, many young people protested the Vietnam War.

This is the only situation when apostrophes mean "more than one."

Exercise 24: Apostrophes

Instructions: Insert apostrophes where needed. Hint: Remember that apostrophes are used in "of" ideas. They don't signify "more than one." When you're finished, check your answers on page 216.

1. The sergeants desk is cluttered with papers.

2. Her stepchildrens claims are unfounded.

3. The puppies were turned over to an animal shelter.

4. The puppies condition is expected to improve.

5. Miss Jones office is down the hall.

6. We all benefited from hearing James explain the new policy.

7. James explanation cleared up several misunderstandings.

8. Families need to understand the special nature of police work.

9. Both instructors did an excellent job.

10. After a weeks vacation, I was ready to return to work.

Exercise 25 More Practice with Apostrophes

Instructions: Insert apostrophes where needed. Not every sentence needs an apostrophe. Check your answers on page 217.

1. I dont understand how to use this calculator.

2. Once again, the repairs to the Smiths' house didn't pass inspection.

3. Two days work was lost when the computer system went down.

4. Lieutenant Conner asked me to address the familys concerns.

5. Last months paychecks will be ready at nine o'clock.

6. I saw the permit on the front door of the Browns house.

7. Lewis inspection was thorough and efficient.

8. The Browns were out of town all weekend.

9. The smoke alarm in the childrens bedroom is missing.

10. Mrs. Hansens contractor applied for the permit last Friday.

Chapter 25

Quotation Marks

On page 125 you learned about quoting victims, witnesses, and suspects accurately. In this chapter you'll learn how to use periods and commas with quotation marks. There are two basic principles to remember:

1. In the United States, periods and commas always go *inside* (before) quotation marks at the end of a sentence. There are no exceptions. (Canada and the United Kingdom use a different system.)

Note these examples:

Linda said, "I checked the fire alarm panel. The power was off."

"Move the trusses out of the emergency access road," I told Wallace.

"When did you arrive at the job site?" Lewis asked.

Lane said, "Final or no final, I'm moving in."

Patel tried to open the door marked "Emergencies only."

"Stop!" I shouted as the electrician's helper reached for the panel box.

2. Use quotation marks only for a person's *exact* words. If you change the words in any way, omit the quotation marks.

Linda said, "I checked the fire alarm panel. The power was off." QUOTATION MARKS

Linda told me she the fire alarm panel and the power was off. NO QUOTATION MARKS

I asked Potter, "Do you know the engineer of record?" QUOTATION MARKS

I asked Potter if he knew the engineer of record. NO QUOTATION MARKS

Exercise 26 Using Quotation Marks

Instructions: Make any corrections that are needed in these sentences. Use the sentences on page 125 as models. (Some sentences may already be correct.) When you're finished, check your answers on page 217.

1. Katherine said that "she had never spoken to her landlord, and she had mailed the rent check to Brooklyn."

2. Sarah told me, "I heard nail guns and knew they were working upstairs without a permit".

3. Brent said, "My electrician went to the supply house."

4. I asked, did you call for a final inspection?

5. "I asked a neighbor if I could look at the house from her back yard." Said Barton.

6. "The electrical system can't carry that air-conditioner", said Farrell.

7. Inspector Hoffman warned Rogers, "not to stock the store before his sprinkler final."

8. "When will you complete your code class?" I asked Susan.

9. Inspector West asked Linda Hamilton if she had heard any strange noises before the deck collapsed.

10. Call an ambulance. Someone fell from the ladder I shouted.

Chapter 26

Pronouns

Pronouns are short, everyday words like *I, me, we, us, she, he,* her, *him, they, us, you,* and *it* that we use in place of other words. It would be clumsy to say something like "Mary said that Mary can't come to the meeting." Most of us prefer to say, "Mary said that she can't come to the meeting."

In most sentences it's easy to use pronouns correctly. But there are four pronoun issues that every professional should know. You'll be reviewing them in this chapter.

1. Singular Pronouns

Several commonly used pronouns are always singular: *any each every someone somebody everyone everybody anybody nobody*

> Every inspector was on time for the meeting. SINGULAR
>
> Somebody needs to enter these statistics. SINGULAR
>
> Any employee is eligible for the program. SINGULAR

Here's where the confusion arises: In everyday conversation we think of these words as plural. Picture *every officer* in your mind, and you'll probably imagine a room full of men and women in uniform. But a closer look at these sentences indicates that these pronouns are singular:

> Every **inspector** [not *inspectors*] **was** [not *were*] on time for the meeting.
>
> Somebody **needs** [not *need*] to enter these statistics.

Any employee [not *employees*] **is** [not *are*] eligible for the program

Words containing *any*, *one*, or *body* are singular:

Everyone uses a laptop to write reports. CORRECT

Everybody likes our new headquarters. CORRECT

Use singular words with singular pronouns (*his, her, its*).

Everyone should have his or her reports completed. SINGULAR

Somebody needs to do his or her job better. SINGULAR

Any inspector can view his or her evaluation beginning on Monday. SINGULAR

There were many smiles today because every employee is happy about his or her raise. SINGULAR

Many professional writers dislike "his or her," even though it's correct, and they find ways to avoid using it. Often you can revise a sentence to avoid "his or her." One strategy is to make the sentence plural. For example, here's a sentence you just read that requires "his or her":

There were many smiles today because every employee is happy about his or her raise.

If you make the sentence plural, you can avoid "his or her":

There were many smiles today because employees are happy about their raises. CORRECT *and* more natural

Sometimes you can avoid "his or her" by substituting "a" or "the":

Each inspector will need three copies of his or her time sheet. CORRECT

Each inspector will need three copies of the timesheet. CORRECT *and* more natural

2. Its or It's?

Use *its* (no apostrophe) as a possessive word (similar to *his*):

Every department is making adjustments to **its** proposed budget.

My uniform is missing one of its buttons.

Remember that **it's** (with an apostrophe) always means **it is**:

When the fire alarms sounds, it's time for the residents to evacuate.

Because it's late, I'll make the phone call tomorrow.

3. The "Thumb Rule"

Use the "thumb rule" when a name appears with a personal pronoun: *I, me, she, her, he, him, we, us, they, them.*

Here's how: Make the sentence shorter by covering the *and* phrase with your thumb. Then use your ear to choose the pronoun that sounds right.

Let Jane and (I, me) help you.

Let ~~Jane and~~ me help you.

Let Jane and **me** help you. CORRECT

Yesterday Jane and (I, me) helped Greg.

Yesterday ~~Jane and~~ I helped Greg.

Yesterday Jane and **I** helped Greg. CORRECT

4. Comparisons

In comparisons, "finish the sentence" by adding an extra word: Your ear will tell you which pronoun is correct. (Go to page 147 to learn more about comparisons.)

Bill is older than (I, me).
Bill is older than I [*am*].
Bill is older than I. CORRECT

Cheryl works faster than (he, him).
Cheryl works faster than he [*does*].
Cheryl works faster than **he**. CORRECT

Joe speaks Spanish better than (I, me).
Joe speaks Spanish better than I [*do*].
Joe speaks Spanish better than I. CORRECT

Carole has been with the agency almost as long as (we, us).
Carole has been with the agency almost as long as we [*have*].
Carole has been with the agency almost as long as **we**. CORRECT

Exercise 27 Pronouns

Instructions: Make corrections in the sentences below. Not every sentence needs corrections. When you're finished, check your answers on page 218.

1. Did everyone complete their requirements for inspector certification?

2. Its obvious that the academy needs to revise its curriculum.

3. Jill has more confidence taking tests than me.

4. Implementing the new policy is going to be difficult for the captain and I.

5. Everyone in the agency has been talking about their upcoming evaluations.

6. Lois replied to this email before she forwarded it to Mr. Morris and he.

7. Ken understands the procedure better than her.

8. The agency is proud of its' safety record.

9. Someone didn't sign their timesheet for this month.

10. No one knows that part of town better than her.

Chapter 27

Verbs

Verbs are action words (words like *go, work, help,* and *run*). Most of the time verbs are easy to use correctly. You should be aware, though, of common verb mistakes that can mar your professional image:

1. Using *seen* without a helper (*is, are, was, were, has, have, had*):

Caruthers seen him with his sister several times. INCORRECT

Caruthers had seen him with his sister several times. CORRECT

Caruthers saw him with his sister several times. CORRECT

Using *done* without a helper:

Hossain done time for burglary in Tennessee. INCORRECT

Hossain had done time for burglary in Tennessee. CORRECT

Hossain did time for burglary in Tennessee. CORRECT

2. Using *snuck* (considered slang) instead of *sneaked*:

Chan snuck into the closet outside the major's office and stole a box of pens. INCORRECT

Chan sneaked into the closet outside the major's office and stole a box of pens. CORRECT

3. Placing the apostrophe in the wrong place in contractions: Remember that the apostrophe takes the place of a missing letter: For example, *do not* becomes *don't*; *is not* becomes *isn't*; *was not* becomes *wasn't*; *I am* becomes *I'm*.

Inspector Farris was'nt on duty yesterday. INCORRECT

Inspector Farris wasn't on duty yesterday. CORRECT

I'am thinking about getting a bachelor's degree in code administration. INCORRECT

I'm thinking about getting a bachelor's degree in code administration. CORRECT

If you're typing on a computer, the spellchecker or grammar checker may warn you that you've made an error. Always check your reports before you submit them, and—if possible—ask a friend or co-worker to read your reports as well. It's much better to catch and correct errors before a supervisor, newspaper reporter, or attorney sees your report.

Verb Endings

Many people have difficulty with *–s* and *–ed* verb endings, especially during conversation. When people talk, they naturally run sounds together, and we tend to omit letters. In most conversations, that's not a problem. But those omitted letters will detract from the professionalism of a report you're writing.

For example, listen to yourself read this sentence aloud:

Bill tried to find the source of the fault current. CORRECT

Chances are you ran the *d* in *tried* together with the *t* in *to*—that's what most people do.

Here's the problem, though: Are you going to remember to *write* that *-ed* ending, since you don't hear or say it? All too often, officers write sentences like this:

Bill try to find the source of the fault current. INCORRECT

Here's another one. Again, listen to yourself read this sentence aloud:

The memo lists the days and times for next month's meetings. CORRECT

Chances are you omitted the final "s" in "lists": It's a difficult word to pronounce correctly, especially when you're talking fast. As a result, the sentence may look like this when an officer writes it:

The memo list the days and times for next month's meetings. INCORRECT

Memo is singular, so you should write *lists* in this sentence:

The memo lists the days and times for next month's meetings. INCORRECT

A similar problem arises with *supposed to* and *used to*: Many people omit the *–ed* ending.

I use to work every holiday. INCORRECT

I used to work every holiday. CORRECT

We're suppose to receive a raise next month. INCORRECT

We're supposed to receive a raise next month. CORRECT

Wilson use to fix cars before he was hired. INCORRECT

Wilson used to fix cars before he was hired. CORRECT

We're suppose to attend a training session next Tuesday. INCORRECT

We're supposed to attend a training session next Tuesday. CORRECT

Adding Verb Endings

Misspellings often creep in when writers add endings to verbs. You can avoid most errors by following a few simple rules:

1. In general, drop the silent *e* when you add a verb ending that starts with a vowel:

   state   hope   care

   stating   hoping   caring

2. Keep the silent *e* when you add a verb ending that starts with a consonant:

   state   hope   care

   statement   hopeful   careless

3. When you're adding an ending to a word that ends with *y*, change the *y* to *i* when it is preceded by a consonant.

   supply   worry

   supplies   worries

4. *Don't* drop the final *y* when you're adding *-ing*.

   study   carry

   studying   carrying

5. *Don't* drop the final *y* when it's preceded by a vowel.

   obey   say

   obeying   saying

Exercise 28 Verbs

Instructions: Correct the verb errors in these sentences. Not every sentence needs corrections. When you're finished, go to page 219 to check your answers.

1. We use to write all our reports by hand.

2. Inspector Larsen did'nt see the memo about the new inspection procedure.

3. Pollard said she seen the contractor last Friday on the job site.

4. I snuck Catherine a piece of candy during the meeting.

5. The notice of violation consist of three items and a correction order.

6. It's going to take a while for me to get use to the new administrative search warrant forms.

7. Perkins is suppose to be graduating from the code academy tomorrow.

8. Inmates from the vocational program done most of the interior work on the new administration building.

9. The report list everyone who has a doctor's appointment today.

10. I've been studing so hard for this exam that I'am sure I'll pass with flying colors.

Chapter 28

Subject-Verb Agreement

Subjects and verbs are the basic building blocks of sentences. These rules will help you avoid mistakes subjects and verbs in any sentence you write.

1. When a sentence begins with *there* or *here*, reverse the sentence to get the verb right.

Here (is, are) your assignment.

THINK: Your assignment **is** here.

Here **is** your assignment. CORRECT

There (go, goes) two fine inspectors.

THINK: Two fine inspectors **go** there.

There **go** two fine inspectors. CORRECT

Here (come, comes) trouble.

THINK: Trouble **comes** here.

Here **comes** trouble. CORRECT

There (seem, seems) to be many possibilities.

THINK: Many possibilities **seem** to be there.

There **seem** to be many possibilities. CORRECT

2. Don't be fooled by numbers. A *unit* of time or measurement is always singular.

Twenty minutes **is** usually sufficient for interviewing an applicant. (unit of time—singular)
Twenty inspectors **are** taking exams for promotions. (twenty separate inspectors —plural)

Two suspects **are** waiting to be interviewed. (two separate applicants —plural)
Two days **is** barely enough time to catch up on my sleep during the weekend. (unit of time—singular)

Five feet **is** the average distance between the tables in the restaurant. (unit of measurement—singular)
Five cars **are** parked illegally on Main Street. (five separate cars—plural)

3. In *either/or, neither/nor* sentences, use the words near *or/nor* to choose your verb.

Neither the contractors **nor the inspector likes** the new regulations about permits. CORRECT
Neither the inspector **nor the contractors like** the new regulations about permits. CORRECT

Neither your report **nor the newspaper articles have** the right information. CORRECT
Neither the newspaper articles **nor your report has** the right information. CORRECT

4. Remember that prepositions (*in, by, for, with, to, of*) introduce phrases that must be crossed out before you choose the verb. (See pages 151-2.)

One ~~of the lockers~~ **is** empty. (skip "of the lockers")

The box ~~on the top shelf~~ is heavy. (skip "on the top shelf")

The inspectors ~~in the Building Department~~ are getting new computers today. (skip "in the Building Department")

5. Words like *each, every, any, everybody, anybody* are always singular.

Each of the witnesses **is [not are]** telling a different story. (*Each* means *Each one*—singular)

Every inspector **has [not have]** a work assignment. (*Every inspector*—singular)

Everyone from both departments **was [not were]** here for the meeting. (Look for the singular word "*one*" in "*everyone*")

Notice that *somebody* contains the singular word *body*—and so does *anybody*.

Exercise 29 Subject-Verb Agreement

Instructions: Choose the correct word in each sentence. When you're finished, go to page 219 to check your answers.

1. Twenty minutes (isn't, aren't) long enough to fill out the form correctly.

2. One of the windows (wasn't, weren't) locked.

3. Neither the brakes nor the clutch (seems, seem) to be working properly.

4. Either the street valve is closed or water main (is, are) obstructed by debris.

5. Advertising for new positions (is, are) going to be posted tomorrow.

6. Departmental policy about interviews (needs, need) to be reviewed by an attorney.

7. Each of the subcontractors (is, are) telling us a slightly different story.

8. All of the subcontractors (is, are) in agreement on some of the details, however.

9. There (is, are) problems with Praeger's site plan.

10. There (is, are) a good reason why the Building Commissioner has doubts about this project.

Chapter 29

Capital Letters

Capital letters aren't difficult to use correctly. Most people know about capitalizing personal names, months of the year, days of the week, and place names. A few special rules sometimes cause difficulty, however. This chapter will help you master those rules (they're surprisingly simple, when you think about them) and use capital letters confidently.

Capital letters present special challenges today because texting is so popular. It's easy (and a serious mistake!) to fall into texting practices when you're typing a report. Always capitalize *I* and the names of people and places, and be sure to apply all the rules in this chapter in every job-related writing task.

Capital Letters Made Simple

1. Capitalize words like *North*, *South*, and so on only when they refer to specific parts of a nation: Midwest, Northeast, the South, the Deep South, and so on. Use lower case the rest of the time.

> I grew up in the Northeast but moved to the Midwest after I married.
>
> Several fires have occurred in the northern sections of town.

2. Capitalize anything that might appear on a sign. Otherwise, use lower case.

> My sister is away at college until November.
>
> My sister attends Florida Southern College.

I think you need to go to the hospital.

North Shore Hospital has an excellent reputation.

The shop at the corner of First and Broadway was robbed last night.

After my last class, I often have a snack at Sam's Snack Shop.

3. Capitalize days and months, but not seasons.

Every winter our workload decreases as the pace of construction slows.

December usually sees an increase in frozen pipes.

Every Tuesday we have a staff meeting.

4. *Always* capitalize languages.

Inspector Perez grew up speaking both English and Spanish.

Does anyone here speak Arabic?

5. Don't capitalize other academic subjects unless they're part of the title of a course, and don't capitalize careers.

I enjoyed biology in high school, but I didn't like physics.

During Building Construction 101, I became aware of the amazing range of career possibilities in the construction field.

6. Capitalize words like *Mother, Father, Aunt,* and *Pastor* only when they're used as people's names.

Did you talk to Mother about her plans for the weekend?

My mother is proud that I chose a career as a building inspector.

We invited Pastor Taylor and Rabbi Levine to the committee meeting about housing the homeless.

We invited both a rabbi and a priest to give invocations.

Exercise 30 Capital Letters

Instructions: Make corrections where needed in the sentences below. When you're finished, check your answers on page 220.

  1. My mother, father, and grandfather proudly attended my graduation from north central police academy two years ago.

  2. Although english and science have never been easy for me, I'm thinking of enrolling in college this fall.

  3. The professors who teach Building Trades courses have an excellent reputation.

  4. You'll enjoy taking Materials Science I and II with professor Henry.

  5. If you're not sure about a career, you should investigate the possibilities in engineering analysis and materials testing.

  6. I'm seriously thinking about becoming an Engineer, and my sister plans to become an Architect.

  7. I first became interested in code enforcement police work when an elementary school teacher taught our class about building inspections.

  8. We meet weekly during the school year and did special projects in the Summer when school was out.

  9. Although Mr. West is retired, the staff at Tracy elementary school still displays the safety posters he made.

  10. Many inspectors received their first introduction to code enforcement working in the trades.

Chapter 30

Comparisons

"Better than," "as good as," "rather than": These kinds of comparisons sometimes appear in inspection reports. Good writers know there are some pitfalls to watch for when you're using these and similar words to make comparisons.

First, remember that our English language is often concerned with the numbers *two* and *three*:

- Use *-er* comparisons (*better, faster, older,* and similar words) when you're comparing **two** people or things. (The word *worse* and phrases beginning with *more* also fall into this category.)

- Use *-est* words when you're comparing **three or more** people or things. (The word *best* and phrases beginning with *most* also fall into this category.)

Inspector Morgan is *more* experienced than Inspector Brown. CORRECT (comparing two people)
Inspector Morgan is the *most* experienced inspector on the force. CORRECT (comparing three or more people)

If you'd spent some time riding with Larry and Tom, you'd know that Larry is the better driver. CORRECT (comparing two people)
Larry is the best driver in our agency. CORRECT (comparing three or more people)

Next, be sure to use *than* (not *then*) in comparisons.

I'd rather work on Saturday than Sunday. CORRECT

The coffee from the staff canteen is better than the coffee in the mess hall. CORRECT

Alan is usually more thorough than she. CORRECT

Finally, when you're writing a comparison sentence, pay extra attention to pronouns (*he, she, I, we,* and so on). Take a look at the last example. Many people would (incorrectly) write it this way:

Alan is usually more thorough than **her**. INCORRECT

If you add an extra word ("is," in this sentence), you can hear that **she** is needed:

Alan is usually more thorough than she **is**. CORRECT
Alan is usually more thorough than **she**. CORRECT

By keeping these pointers in mind, you can handle comparisons effectively every time. (To review using pronouns in comparisons, go to page 130.)

Exercise 31 Comparisons

Instructions: Choose the correct word in each sentence below. When you're finished, check your answers on page 221.

1. I'd rather inspect a service station (than, then) work in the office.

2. Margaret is nearly as good at report writing as (he, him).

3. Brock is the (best, better) of the two drivers.

4. Out of all the places I've worked, I like this agency (best, better).

5. Few people face as many complaints as (we, us) in code enforcement.

6. Calvin is the (worse, worst) driver in the department.

7. I'm good at setting up spreadsheets in Excel, and Gary knows almost as much as (I, me).

8. I like outdoor work much more (than, then) sitting in an office cubicle.

9. I tried both laptops, and this one is definitely (better, best).

10. Which of the three applicants is (more, most) qualified?

Chapter 31

Prepositions

The grammatical term *prepositions* sounds intimidating to many people. But it doesn't have to be. The simple truth is that you've been using prepositions ever since you learned how to speak...and you've probably used them correctly most of the time. As a serious writer you need to learn only a few usage rules about prepositions.

What are prepositions? They are small, ordinary words that indicate direction or purpose: *in, by, for, with, to, of, on, over, under, beside, near, along*...you can probably think of many more.

Prepositional phrases are small word groups that begin with prepositions: *in the garden, by the sea, for a year, with my sister, to the store*, and so on.

Here are the usage points you need to know:

1. Most of the time prepositional phrases are **extra** parts of sentences. When you're analyzing a sentence, you should usually skip over the prepositional phrase to get to the really important parts.

> A change in city policies are causing headaches for construction inspectors. INCORRECT

What is the sentence really about? Answer: A change. "City policies" aren't causing the headaches: The *change* is.

So the sentence needs to be corrected:

A change in city policies **is** causing headaches for construction inspectors CORRECT

(You can learn more by reading about Rule 4 on page 140.)

2. You can use a comma when a sentence begins a prepositional phrase. Note, though, that many good writers omit the comma if the prepositional phrase is short. It's your choice.

> On Tuesdays the building commissioner meets with the mayor. [No comma: *On Tuesdays* is a short prepositional phrase.]

> Under the subflooring in the apartment next door, I found black mold on the joists. [Use a comma: *Under the subflooring in the apartment next door* is a long prepositional phrase.]

You can learn more about these commas by reading about Comma Rule 1 beginning on page 105.

3. Shorten the sentence and use your ear when a pronoun (*he*, *she*, *him*, *her*, *I*, *me*, etc.) follows a preposition.

> I gave the report to her for proofreading. CORRECT [not "to she"]
> I gave the report to Fakir and **her** before I delivered it to the mayor. CORRECT [not "to Fakir and she"]

> Commissioner Strong thanked me for my hard work. CORRECT
> Commissioner Strong thanked Inspector Brown and **me** for my hard work. CORRECT

You can learn more about sentences like these by reading about the "Thumb Rule" on page 129.

4. Use prepositions with precision. Notice the different meanings in these two sentences:

> Inspector McCaffrey walked in the room. [He spent time walking around the room.] CORRECT

Inspector McCaffrey walked into the room. [He entered the room.]
CORRECT

Exercise 32 Prepositions

Instructions: Correct the errors in the sentences below. Not all sentences contain errors. When you're finished, check your answers on page 222.

1. In an apartment on the second floor I found two cross-connections.

2. An array of arc fault circuit interrupters, ground fault circuit interrupters, and tamper-resistant receptacles was displayed on a table at our booth at the fair.

3. A decision about criminal charges for the unlicensed contractors are expected by tomorrow morning.

4. For now the new commissioner is concentrating on getting to know the inspectors and the challenges they're facing.

5. By the end of next October we will be ready to discuss the new budget.

6. Under the sink in the kitchen I found a leaking trap.

7. Misuse of temporary heaters by contractors cause significant property loss.

8. Of all the applicants for the new position Patricia Cooney seems the most qualified.

9. An assortment of undersized fasteners were found when we investigated the porch collapse.

10. For the most part unlicensed contractors don't seem to understand the seriousness of what they've done.

Chapter 32

Avoiding Common Errors

This chapter is a refresher about usage points that cause problems for many writers. Take a look at these quick, easy-to-understand explanations. In just a few minutes you can clear up some of the most common mistakes found in code enforcement reports.

1. Not ending sentences with a period.

Here's a quick lesson: Extra ideas end with commas. Sentences end with periods. (See page 81 for more help.)

When McKay asked for a GFCI tester, EXTRA IDEA

McKay asked for a GFCI tester. SENTENCE

Although I was on leave, EXTRA IDEA

I was going on leave. SENTENCE

Here's how to put extra ideas and sentences together:

When McKay asked for a GFCI tester, Jackson responded. CORRECT

Although I was on leave, I decided to check on McKay. CORRECT

2. Not knowing what a sentence is.

Here's a quick lesson: A sentence begins with a person, place, or thing.

He ran out the door. SENTENCE

After he ran out the door, EXTRA IDEA

After he ran out the door, Clara called 911. CORRECT

It's a sentence even if it's short or unclear:

I understand. SENTENCE

He did. SENTENCE

It is here. SENTENCE

Go to page 97 for more help with periods, and to page 103 for more help with commas.

3. Using a comma instead of a period to start a new sentence with *it*. Good writers use a period to start a new sentence.

The rope broke, it wasn't strong enough. INCORRECT
The rope broke. It wasn't strong enough. CORRECT

I rejected her explanation, it didn't make sense. INCORRECT
I rejected her explanation. It didn't make sense. CORRECT

4. Getting pronouns mixed up (*I/me, he/him, she/her, we/us, they/them*).

Jim and me inspected the warehouse. INCORRECT (Think: I inspected the warehouse.)
Jim and **I** inspected the warehouse. CORRECT

The commissioner gave Cynthia and I a special assignment. INCORRECT (Think: The commissioner gave **me** a special assignment.)
The commissioner gave Cynthia and **me** a special assignment. CORRECT

I spoke to he and his wife about the procedure for obtaining a permit. INCORRECT. (Think: I spoke to **him**.)
I spoke to him and his wife about the procedure for obtaining a permit. CORRECT.

For more help with these pronouns, go to page 129 and read about the Thumb Rule.

5. Using unnecessary apostrophes with the letter *s*.

Here's a quick lesson: Use apostrophes only in contractions (*don't, can't*) and "of" ideas (*Mary's car, Tom's schedule*). Apostrophes aren't decorations, and they don't mean more than one.

> The cars will be replaced in two years. (no "of" ideas, and no apostrophes) CORRECT
>
> John's laptop isn't working properly. (laptop of John) CORRECT
>
> The Browns are taking a vacation next month. (no "of" ideas, and no apostrophes) CORRECT
>
> My uniform doesn't need to be dry cleaned until next week. (*don't* is a contraction) CORRECT

Go to page 119 for a review of apostrophes.

6. Putting a comma after a subordinate or coordinate conjunction.

Sounds intimidating! But actually those conjunctions are words you use every day: *and, but, if, when, because, although,* and so on. Don't put commas after them. (If you need a comma, put it in front.)

> I walked around the whole perimeter of the store but, I didn't see or hear anything. INCORRECT
> I walked around the whole perimeter of the store, but I didn't see or hear anything. CORRECT
> I walked around the whole perimeter of the store but didn't see or hear anything. CORRECT

Be especially careful with *although*. Anything that starts with *although* is an extra idea that has to be attached to a real sentence (and of course you'll never put a comma after *although*):

> He insisted on driving his car home. Although, his friends tried to stop him. INCORRECT
> He insisted on driving his car home although his friends tried to stop him. CORRECT

For more help with Comma Rule 1, go to page 105.

7. Misspelling *all right* and *a lot*.

All right and *a lot* are always two words. Always. There are no exceptions. You can check the dictionary to verify this: It will tell you that the common one-word spellings are "nonstandard," meaning that professionals never use them. See also page 179.

> Leon told me he was all right and didn't need medical attention. CORRECT

> Denise said she'd heard a lot of yelling coming from the Wrights' apartment. CORRECT

8. Misusing quotation marks.

Use quotation marks *only* for a person's *exact* words. If you change the words, omit the quotation marks. Go to page 125 for more help.

When you use quotation marks, always put commas and periods inside. There are no exceptions in the United States. Go to pages 125-6 for more help.

> Casey said, "I was afraid the wall would collapse, so I ordered the workers to evacuate." CORRECT

> Casey said she was afraid the wall would collapse, so she ordered the workers to evacuate. CORRECT

9. Using texting style.

Because texting is so popular, many people have become careless about abbreviations and capital letters. Beware! If you text often, ask someone to check your reports to make sure you haven't slipped into texting style.

I will call u when i no the location for our September meeting. INCORRECT

I will call you when I know the location for our September meeting. CORRECT

10. Misusing verbs or forgetting to use "helping verbs" like *is, are, was, were, has, have,* and *had.*

Writing the way you speak can cause huge problems. Be especially careful with commonly misused verbs like *seen, went, did,* and *done.*

My partner and I seen her get into her car. INCORRECT
My partner and I saw her get into her car. CORRECT

After I had went to the parking garage, Thompson changed his story. INCORRECT
After I had gone to the parking garage, Thompson changed his story. CORRECT

Wilkes said he done everything by four o'clock. INCORRECT
Wilkes said he did everything by four o'clock. CORRECT
Wilkes said he had done everything by four o'clock. CORRECT

Exercise 33 Avoiding Common Errors

Instructions: Correct any errors in the sentences below. When you're finished, check your answers on page 222.

1. When I seen smoke coming from the engine, I called the fire department.

2. The lock is sticking, it probably needs some graphite.

3. Alot of inspectors are watching the weather closely because, the US Weather Service is predicting a hurricane.

4. I don't no the answer to that question but i will find the answer for you.

5. Many people think that crime rates go up every year but, in many places crime rates have gone down.

6. Although I gave the data to Mary Alice was the one who entered it into the database.

7. When I questioned Fischer about the batteries', he said, "he didn't know how they got there."

8. Later Alan Jencks, Fischer's supervisor, said "I don't allow any workers to remove batteries from the shop".

9. We done everything we could to prepare our headquarters for the open house next Saturday.

10. I'm signing up for a Spanish course next month although, French might be more useful in some neighborhoods.

Chapter 33

Myths about Grammar

There's a lot of misinformation about writing out there. Most people, including code inspectors, learned their writing skills in elementary school from teachers who weren't professional writers or editors. Myths that started decades ago continue to cause endless confusion. Let's clear up some of them.

Discard These Writing Myths

1. "A comma takes the place of *and*."

No, it doesn't. You won't find this made-up rule in any grammar book. Often, in fact, you need a comma with *and*. (Suggestion: Review Comma Rule 2 beginning on page 109.)

What if you have a list of three items—do you use a comma before *and*?

The answer is that it varies (except in journalism, when it must be omitted.) You can do it either way as long as you're consistent—or you can ask what your supervisor prefers. Many writers find that the comma makes sentences easier to read, so they always use it in a series.

> I inspected the furnace, red-tagged it, and revoked the permit. CORRECT
> I inspected the furnace, red-tagged it and revoked the permit. ALSO CORRECT.

2. "*Ain't* ain't in the dictionary."

Yes, it is—and it always has been. (Look it up!) *Ain't* is a word. Admittedly it's slang, but nevertheless it's a real word with a long history. If you're writing down a witness's exact words, and the witness uses *ain't*, go ahead and include it in your report. Otherwise avoid *ain't*: It can damage your professional image.

3. "You can't start a sentence with *but*" (or *and* or *because*).

There's no such rule. (Go to the library and look for it! It doesn't exist.) Good writers start sentences with these words all the time. Pull some books off your bookshelf and see for yourself. Check the newspapers and magazines you read regularly.

4. "Don't use *I* or *you* when you write."

Old-time report-writing teachers used to warn against "I" and "you" because they were worried about objectivity. What we've learned, of course, is that crossing out "I" and writing "this inspector" accomplishes...nothing.

Once again, thumb through some books and magazines and see for yourself. You'll find that good writers use "I" and me" all the time. The prohibition applies only when you're writing for an academic publication or something else that's very formal, such as a legal contract. (To learn more, turn to page 60.)

5. "Use a comma with a person's name."

Not true. Use the comma with a name only when you write a Comma Rule 3 sentence, like this:

> Patricia Gavin, who joined the agency last week, graduated from the academy with high honors. CORRECT

Don't use a comma in other sentences containing a name:

> Patricia Gavin has dreamed of a career in code enforcement since she was ten. CORRECT

(To learn more about Comma Rule 3, turn to page 113.)

6. "Use a comma when you pause."

True professionals use the three comma rules to ensure their sentences are correct. (See page 103.) If you want to pause for emphasis, use an ellipsis (three periods, like this: …)

> When Commissioner Corey retired, we surprised him with…a trip to Hawaii. CORRECT

> On a hunch, I opened the panel and saw…a rejected sticker from two years ago. CORRECT

Save these special effects for writing projects away from your code enforcement job (such as college papers, personal letters, and writing for publication). Avoid trying for special effects when you're writing a report.

Section V

Choosing the Correct Word

Chapter 34

Terminology

Like other professional fields, code administration and enforcement has its own specialized vocabulary. Most are found in the technical codes, but some words are standard legal terminology. Terms vary from code to code and state to state. It's important to know the meanings of these words and to use this vocabulary precisely.

This chapter reviews words every inspector should know. (Because of variations in local laws and procedures, exact definitions may be different in your jurisdiction.) The International Code Council's "I" Codes, dedicate Chapter Two of each technical code to definitions. The National Fire Protection Association generally dedicates Chapter Three to definitions, although definitions in NFPA 70, *National Electrical Code* are in Chapter One.

Adopt To accept and formally put into effect.

Approved Acceptable to the code official with jurisdiction. Approval is normally based on nationally recognized standards, or in their absence, on sound engineering practice.

Citation A written order, issued by a law enforcement officer or other authorized official, directing an alleged offender to appear in court at a specific time to answer a criminal charge. Civil citations are used in some states for minor infractions and code violations.

Closed or **closed down** Inspectors sometimes describe their action in ordering the evacuation of an unsafe structure as "closing it down." Unless your code gives you the authority to *close* a business, do not use this term. The appearance of exceeding your authority can be as damaging to you and your agency's reputation as actually exceeding it.

Freedom of Information Act or FOIA Federal and state laws that ensure public access to public records and to the meetings of deliberative bodies of government, also called *sunshine laws*. See Chapter 38.

May Indicates a discretionary provision of the code. Enforcement is left to the judgment of the code official.

Selective Enforcement Code enforcement programs that target select areas, types of businesses, or groups of residents. Selective enforcement based on race, religion, national origin or sexual preference is illegal, but selective enforcement based on actual hazards, fire frequency, or other emergency incidents is not.

Shall Indicates a positive and definitive requirement of the code that must be performed. Action is mandatory.

Summons A written order issued by a judicial officer, law enforcement officer, or other authorized official, directing an alleged offender to appear in court to answer a criminal charge. A summons issued by a law enforcement officer is classified as a citation.

Warrant A legal writ issued by a judicial officer commanding an officer to arrest a person, seize property, or search premises.

Terms to Avoid

It's important that your reports and notices are easy to read and understand. Avoid using common slang and slang terms commonly used in the trades. Common slang makes you look unprofessional. When you use industry slang, electricians, carpenters, and plumbers might know what you mean, but hearing officers, judges, the press, and the public probably won't.

Chapter 35

Words and Expressions to Avoid

Inspectors often use jargon when speaking and writing. Sometimes it's to save time: CO (certificate of occupancy), NOV (notice of violation), OSB (oriented strand board). Sometimes we use the brand name of a specific product to describe a material produced by all manufacturers: Celotex™ (black fibrous board exterior sheathing), Xerox (photocopier). It's natural to want to sound like other inspectors, but caution is sometimes necessary.

But there are several good reasons why you should avoid certain outdated expressions and jargon in your inspection reports: They're not professional, they might confuse outsiders who read your reports, and the jargon habit can be hard to break when you go on to other forms of writing (if you're promoted, for example).

Abovementioned

Outdated and often unnecessary. Either omit it or replace it with a specific word or name: *Amy Kallen* rather than "the abovementioned complainant."

Ascertained

Another outdated word.

Be specific about how you found the information (especially important for appeal hearing or court testimony): *saw*, *heard*, *read*, and so on. Instead of writing "I ascertained that the homeowner had installed the chimney without a permit," describe how you determined the fact: a search of the permit database came up empty.

Affirmative

Substitute *yes*.

Approximately

Substitute *about*.

At the present time

This phrase is often unnecessary and can be omitted—or you can substitute *now*.

Being that

Substitute *because*.

Contacted

Not specific enough for court. Substitute *met with*, *talked to*, *telephoned*, *emailed*, *questioned*, *wrote*, or a similar word.

Detected

Substitute *saw, smelled, touched, heard, uncovered*—whatever you did to get the information.

Endeavor

Substitute *try*.

Expedite

Substitute *hurry* or *speed up*.

Fireman or firemen

Substitute *firefighter or firefighters*.

If and when

Substitute *if*, which covers both words.

In close proximity to

Substitute *near*.

In reference to

Substitute *about*.

Modify

Substitute *change*.

Negative

Substitute *no*.

Numerous

Substitute *many*.

Pertains to

Substitute *is about*.

Policeman

Substitute *police officer* or *law-enforcement officer*.

Prison guard

Correctional officer is the proper term for an officer in a jail or prison.

"Processed the area"

This vague phrase should be replaced with a specific description of what you did: "recovered two concrete samples from the floor slab."

Proceeded

Too vague for a report: Did you drive, walk, run, ride a bicycle, ride a Segway? *Proceeded to* can be changed to "entered" or "arrived at."

Render

Substitute *send*.

Residence

Too vague for a report. Be specific: Was it a doublewide mobile home, a house, an apartment, or a condo?

Respective

An old-fashioned word that's almost always unnecessary. See page 61.

Take cognizance of

Substitute *recognize*.

Advised or Told?

The word *advise* is a prime example of jargon that can create problems. Many inspectors mistakenly use *advise* as a synonym for *tell*:

> Barlow advised me that he'd been at work when the hood suppression system operated.   INCORRECT

> Barlow told me that he'd been at work when the hood suppression system operated. CORRECT

This habit is so entrenched in report writing that most inspectors will understand what you mean if you substitute *advise* for *tell* occasionally. But consider what happens if you use *advise* incorrectly in other settings.

For example, suppose you're writing a research paper for college, an article for a professional journal, a press release for a local newspaper, or a supervisory report. The person who reads what you've written is going to be puzzled if you give the impression that advice was given when actually nothing of that sort happened.

> Several teachers at Middleton Junior High advised me that they appreciated my presence at the school. INCORRECT (no advice was involved)
>
> I advised the teachers at Middleton Junior High about fire safety during the school renovation. CORRECT (genuine advice)
>
> Commissioner Cooper advised me that heavy rain was expected. INCORRECT (no advice was involved)
>
> Commissioner Cooper advised me to avoid Lincoln Boulevard, which floods after a heavy rain. CORRECT (genuine advice)

This example uses both *told* and *advised* correctly:

> I *told* Johnson that his electrician's license would expire at the end of the month. I *advised* him to make an appointment right away. CORRECT (only the second sentence involves advice)

Exercise 34 Advised or Told?

Instructions: Change *advised* to *told* where necessary. Answers appear on page 223.

1. I advised Inspector Jones that he was assigned to the morning shift.

2. I advised Inspector Jones to improve his negative attitude.

3. I advised Mary Smith to see a doctor about the cut on her arm.

4. Smith advised me that she cut her arm on the paper cutter.

5. Commissioner Simmons advised us that he would be on vacation the first half of July.

6. Inspector Donaldson's doctor advised him to limit his cholesterol intake.

7. I already advised the building manager about the broken alarm on the third floor.

8. I'm glad I listened to Mr. Johnson when he advised me to continue my education right after high school.

9. The chaplain advised us that there would be a special religious service Sunday evening.

10. My guidance counselor in high school advised me to take a keyboarding course.

Chapter 36

More Words to Watch

A/An

When you're thinking about *a/an*, go by the sound, not the spelling. "A uniform" is like "a youthful offender." "A uniform" is not like "an uncle."

He was wearing a uniform issued by the Red Robin Casino.

He became an uncle when his sister's daughter was born.

Advice/Advise

Advice is a noun (a thing). *Advise* is an action.

I often asked Inspector Jones for advice when I was new to the department.

I would advise you to get medical attention for those bruises.

Do not use *advise* when you mean *tell*. See page 174.

All ready/Already

All ready means "all prepared." *Already* means "by this time."

The lunches are all ready and set up in the mess hall.

We've already been to that apartment building twice.

All right

In the United States, always two words. See page 158.

A lot

In the United States, always two words. See page 158.

Between/Among

Use *between* for two people or things, *among* for three or more.

The hiring committee is trying to decide between two qualified candidates for mechanical inspector.

Cooper, Daniels, and Peterson divided the inspections among themselves.

Break/Brake

Break means "shatter" or "separate"; *brake* refers to stopping and the pedal in a car.

I had to break my appointment to talk with the mayor.

The brake pedal doesn't feel right.

Breath/Breathe

Breath is a noun (a thing). *Breathe* is an action.

I smelled an alcoholic beverage on his breath.

The medic asked her to breathe deeply while he listened to her lungs.

Compliment/Complement

Complement refers to completeness or making something complete. *Compliment* means "praise."

A full complement of officers attended the ceremony.

I want to compliment you on the way you handled that incident.

Comprise/Compose

Comprise means "include." *Compose* means "made up of."

The committee comprised all 15 department heads.

The committee is composed of representatives from every agency.

Curse/Swear

A *curse* is an evil wish; *to curse* is the act of wishing evil upon another. *Swearing* involves calling upon a deity for verification. (Don't confuse *curse* and *swear* with obscene language.)

"Go to hell!" screamed Phillips.

Davis said, "I swear I never touched her."

"You're a bitch!" shouted Tainey when I handed him the notice.

Eminent/Imminent

Eminent means "famous" or "respected." *Imminent* means "about to happen."

She is an eminent authority on DNA.

Seeing no imminent danger, I inspected the second floor

Imply/Infer

Imply means "to hint." *Infer* means "to deduce."

Landers said she struck her husband because he implied she was having an affair.

I inferred that Rogers would be out sick for at least a week.

Individual

Often unnecessary.

Next week ~~individual~~ members will receive their renewal notices.

See page 61.

It's/Its

Its is a possessive word (like *his*) and does not use an apostrophe. *It's* means *it is*. The apostrophe replaces the missing *i*.

> Her car is overdue for its oil change.

> I have to hurry because it's almost time for our meeting.

Never put an apostrophe after *its*: ~~its'~~. Go to page 129 to learn more about *it's* and *its*.

Less/Fewer

Use *fewer* when you're comparing things you can count, like fasteners or victims. Use *less* when you're comparing amounts that can't be counted, like noise. You should also use *less* in expressions like *less than one*.

> We received fewer noise complaints this weekend than we were expecting. ["Noise complaints" can be counted.]

> This weekend there was less flooding than we were expecting. ["Less flooding" can't be counted.]

Lie/Lay

Use *lay* when you place something. Use *lie* (not lay) to refer to resting, sleeping, and napping. (Another way to remember: *Lay* is done to things; *lie* is something you do yourself.)

> Sanders was lying on the sofa, watching a football game, when he heard the water line burst.

> Don't lay anything on that desk; it needs to be dusted first.

Lose/Loose

Loose means "not tight." *Lose* means "misplace."

> When he realized he was going to lose the argument, Felder hit his wife.

Carr was afraid he would lose the money, so he left it in a bureau at home.

More/Most

Use *more* when you compare two things or persons; use *most* to compare three or more.

Code inspectors are more professional today than they were thirty years ago.

I have more experience with concrete construction than Marek has.

Posner is the most skilled block layer on the project.

See page 147.

Of/Have

Don't substitute *of* for *have* when you need a helping verb.

Mattson could ~~of~~ have left through the bedroom window.

Passed/Past

Passed is an action that already happened. *Past* is an adjective referring to a previous time.

Karen said she passed out after taking a few breaths of the gas.

In the past, inspectors wrote their reports by hand or on a typewriter.

Be careful not to write *pasted* ("glued") when you mean *passed*.

Patients/Patience

Patients are people treated by a healthcare professional. *Patience* is the quality of waiting without complaint.

Many of Dr. Morrow's patients came to him for mood-altering prescriptions.

Thorough construction accident investigations require patience and skill.

Personal/Personnel

Personal means "private" or "intimate." *Personnel* are employees.

Jackson kept his personal papers in a locked drawer.

All personnel were asked to stay in the office for a special meeting.

Principal/Principle

A *principal* is the director of a school; *principal* also an amount of money that has been borrowed or invested, and it's an adjective meaning important. A *principle* is a truth, rule, or law.

The principal wants to expand the D.A.R.E. program in her school.

Our principal concern is the possibility of a fire.

Having high principles and sticking to them is vital to the code enforcement profession.

Quiet/Quite

Quiet means "not noisy." *Quite* means "rather" or "very."

It was two a.m. before the dorm was quiet again.

The evidence is quite clear.

Saw/Seen

Use *saw* by itself, and use *seen* with a helping verb (*am, is, are, was, were, has, have, had*). Go to pages 133 and 159 for more help with *saw* and *seen*.

I saw the manager before he spotted me.

We have already seen good results from the new training program.

Stationary/Stationery

Stationary means "not moving." *Stationery* refers to paper products used for correspondence.

I spend 15 minutes on a stationary exercise bicycle every morning.

You can get a copy of the report by submitting a request on official department stationery.

Than/Then

Than is a comparison word; *then* is a time word.

Sometimes a dog's nose is more sensitive than our sophisticated laboratory equipment.

If you're interested in advancement, then you should think about going back to college.

See also page 147.

Their/There/They're

Their refers to ownership by two or more people. *There* is an adverb similar to "here." *They're* is a contraction of "they are."

The junk dealer took all their old parts.

Couture saw step-cracks there and called his masonry contractor.

There are two witnesses waiting to be interviewed about the incident.

They're all in agreement about what happened.

To/Too/Two

To indicates direction or purpose. *Too* means "excessive" or "also." *Two* is a number.

I'm going to the materials lab to submit concrete cores for analysis.

Chen and Wu said they were too frightened to call police.

Horvat is interested in a criminal career too.

I spent two hours searching the database for information about Kovac's business practices.

Who/Whom

Whom is like "him" (notice the final *m* that they both share). *Who* is like "he."

We're trying to discover who had the combination to the safe. [like *he had the combination*]

The officer who did this should be commended. [like *he did this*]

The neighbor whom I talked to gave me a good description of the suspect. [like *talked to* **him**]

Your/You're

Your means "belonging to you." *You're* is a contraction of "you are."

I'm impressed with the thoroughness of your report.

When you're finished, the data will go into our annual report.

Exercise 35 Words Often Confused

Instructions: Make any corrections needed in the sentence below. When you're finished, go to page 224 to check your answers.

1. Were assigned to the committee that's drawing up a uniform policy about time off for work-related travel.

2. Not all agency personal travel to meetings and conferences, so its hard to be fair to everyone.

3. Your going to find that wearing an uniform doesn't automatically earn you respect.

4. If its alright with the chief, we'll be replacing alot of the exercise equipment upstairs.

5. Safety should always be our principle concern.

6. If the report is to long, I'll help you find places to cut it, and Mike can help us to.

7. Officer Campbell's master's in psychology is a nice compliment to her undergraduate degree in law enforcement.

8. That car just past me going sixty in a 30 mph zone.

9. The courtroom became quite when Caruso went to the stand to testify.

10. Whom is going to meet with the city council tomorrow?

Chapter 37

Using Plain English

For code inspectors, the ability to write clear, concise reports and orders is essential. Building and business owners, developers, contractors, and design professionals will take measures to comply with the code based on what you write. Repairs, maintenance, structural changes, and changes in business practices usually have associated costs. A report written in plain English prevents confusion and helps ensure compliance at the least possible cost.

Some inspectors, however, are afraid to use plain English on the job. How can you earn the respect of others with everyday words that everyone knows? Shouldn't you try to show off your knowledge?

The answer is, quite simply, no. You need to impress people with who you are and how you do your job, not with pompous language. This is a universal principle that you can test yourself from your own memory bank. Have you ever known someone who found roundabout, fancy ways to communicate? Were you impressed? Chances are you were simply exasperated.

Now think about people you've had great respect for—family members, other inspectors, friends, and community leaders. What impressed you? Chances are it was their skill and knowledge. In fact you may recall that their vocabularies and sentence structure were pretty ordinary. What caught your attention was their intelligence or some other quality they had developed.

Now think about the people you will be encountering in your code enforcement—citizens, contractors, business owners, design professionals, developers, public officials, as well as their friends and their families. Chances are some will be highly educated and accomplished, while others will not. How important is it for you to communicate clearly?

Now turn your thoughts to the reports you will be writing. There will be lots of them, and they will be time consuming. Do you really want to take the time to write 30 words when 10 or 15 would do the job just as well? Picture yourself preparing for a court appearance by rereading a report you'd written six months earlier. Would you rather read a wordy, complicated report, or one that states the facts in a clear, straightforward way? (See page 52 for more suggestions.)

Writing in plain English has three important advantages for inspectors. First, you can get your reports finished more efficiently. Second, you're less likely to make grammatical mistakes. Most important, the people you're trying to communicate with are likely to understand you the first time. (How many times have you been frustrated by a sentence you had to read several times before you figured out the meaning?)

Choose the Simplest Word

Here's an excerpt from a report about a traffic stop. After you've read it, see if you can find ways to simplify the words and sentences without sacrificing any important details.

> I proceeded to request Ketcham's permit and approved plans. He provided both forms of documentation upon my request. I proceeded to examine them in order to determine whether or not they were in order. Being that the permit application was submitted under the name Carlisle Plumbing, I commenced asking him about the status of his contractor's license.

And here's a simplified version of the same paragraph:

I reviewed Ketcham's permit and approved plans and noticed the application was submitted under the name Carlisle Plumbing, I asked him about the status of his state contractor's license.

Two Ways to Simplify

Good writers use two strategies to achieve clear, straightforward sentences. **First**, they make wise word choices. Instead of "commence," you can use *begin*. "Adjacent to" becomes *next to*, and so on. "Approximately" (often shortened to "approx.") can become *about*. Instead of "apprehending" suspects, you can *arrest* them.

The **second** strategy is to eliminate unneeded words altogether. Compare these example pairs, noticing that no meaning is changed or lost when the highlighted words are removed:

three ~~individual~~ members

three members BETTER

green in color

green BETTER

their ~~respective~~ homes

their homes BETTER

~~At the present time~~ he is a plumbing inspector.

He is a plumbing inspector. BETTER

She ~~currently~~ works as a dispatcher.

She works as a dispatcher. BETTER

He slowly ~~and surely~~ drove to the barn with the roof trusses.

He slowly drove to the barn with the roof trusses. BETTER

The following exercise will help you think about timesaving substitutes for unnecessary words and phrases you hear almost every day. Start listening to yourself and others as you communicate at home and at work. What timesaving changes can you make in your own speaking and writing habits?

Exercise 36 Use Plain English

Instructions: Rewrite this public notice. Think about what's most important, and put that information first. Omit unnecessary words and information. When you're finished, compare what you've written to the simplified version on page 225.

> Building Commissioner Anna Brown and Assistant Commissioner Carl Summers have been looking for opportunities to increase community awareness of services our agency provides to the public. It has been decided to hold a one-day Citizens' Code Academy on Saturday, November 14, from 9 to 4 in the Community Room at the Sixth Street Library. Participants will learn about the department's safety programs and public initiatives. There is no charge. Interested citizens should preregister by calling 555-1212 or clicking the appropriate link at our website at www.SmithvilleCodes.org.

Exercise 37 Write Efficiently

Instructions: Write a simpler version of each word or phrase. Reviewing Chapter 35 (pages 173-8) will be helpful. Suggested answers appear on page 225.

utilize	yellow in color
single-click	large in size
PIN number	in the event that
for the purpose of	if or when
the month of October	preplan

pull-down menu

take cognizance of

lower down

scream and yell

brand new

this inspector

inquire

transport

pursue

abovementioned

commence

relative to

initiate

in order to

residence

ascertain

numerous

at the present time

terminate

being that

finalize

endeavor

by means of

in close proximity to

modify

with reference to

answered in the affirmative

approximately

Chapter 38

Freedom of Information and Public Records Laws

This chapter isn't about how to write. It's a reminder about state and federal Freedom of Information Act laws that allow many people to have access to your reports and records. Your reports are also subject to subpoena in criminal and civil cases. It's a good idea to review your agency's policy on records and your state's statutes regarding public records.

The federal *Freedom of Information Act* (FOIA) of 1967 was the first law to open the workings of government to the public. Average citizens didn't stampede Washington to rifle through file cabinets, but press access to agencies and their information was significantly increased. States followed with their own statutes that ensured access to public records and prohibited deliberation by public officials except in duly advertised public meetings.

What Are Public Records?

Like records of any government agency or entity, records generated by and under the control of inspection agencies are public records. *Public records* is a legal term used by governments and does not necessarily mean the documents are available for public inspection.

Although political subdivisions might differ slightly, the common legal definition of public record from *Black's Law Dictionary* is:

> A record, memorial of some act or transaction, written evidence of something done, or document, considered as either concerning or interesting the public, or open to public inspection.

The management, retention, disclosure, and destruction of public records are prescribed by regulations developed by boards, commissions, state archivists, and state statutes.

The federal laws regarding public records within the federal government are in Title 44 of the *United States Code*. Each state has statutes regarding records management by the state government and the political subdivisions within the state. A quick computer search should quickly help you find the statute for your state. It's safe to say that your inspection reports are public records. What is highly likely (and sometimes comes as a surprise) is that your work-related emails and instant messages are public records too. So are text messages on your government cell phone—even the ones sent to you by friends who aren't government employees. Treat every message as if it will end up in the hands of the news media.

Whether an inspection program is operated as a governmental function or by a private concern, the laws regarding public records are applicable—as long as the program is conducted on behalf of the government. That makes the records *public records*. Everything regarding the record's origin, life, and eventual destruction should be done in accordance with the appropriate law or regulation.

There are exceptions, however. Generally, ongoing criminal investigations, medical records, personnel records, items affecting national security, trade secrets, and attorney-client communications are exempt from disclosure.

How Do Public Records Laws Affect You?

How do these laws and regulations affect you, the inspector? Better questions might be: *How can you comply? And what is the potential impact if you disregard the regulations?*

Failure to comply with these laws can get you into trouble. It can also cast a pall on your reputation as a public official. *If she's so cavalier that she ignored the public records law, what other laws did she ignore?* The public rightly expects public servants to conform to all rules and regulations.

Your reports are much more than mere sheets of paper or digital files. They have a lifespan prescribed by your state archivist or by federal statutes and regulations. State archivists publish retention schedules that mandate retention based on the type of file. Some documents must be retained forever. Usually inspections reports must be maintained for a certain number of years, or for the life of the building. Make sure you're complying with the records management rules and procedures of your agency or department. As long as you follow the rules, a court or disciplinary board may have a problem with the rules, but not with you.

Don't assume that you're protected by common-sense distinctions between public and private documents. Suppose during working hours you use a department-issued pen to scribble a grocery list on a department-issued pad of paper while sitting in your inspection department car. Is that grocery list a public record under the FOIA? There's no definitive answer—but it could be subpoenaed.

A good rule is to avoid mixing personal information with business information. A subpoena to produce *any and all field notes, written observations, and written documents* for a given date might be a problem. When a subpoena asks for *all* the records, that's what you need to provide. If the grocery list is in your daily log pad, it goes with the rest of the records. If it's on a scrap of paper in your pocket, it probably doesn't. Your department's legal counsel will make the call. The court will determine if they got it right.

Exercise 38 Understanding Public Records

Directions: Find the public records statute that applies to code enforcement records in your state. Rewrite the statute guidelines in your own words to show that you understand them. If necessary, check with a supervisor to ensure that you know how to comply with public records requirements in your state.

POST-TEST

Instructions: Every sentence has *at least* one error. Make any corrections needed. When you're finished, check your responses against the Post-Test Answer Key beginning on page 201.

1. Parker tried to intimidate me when I told him I was writing a notice of violation.

2. I asked Santos what happened, and he said that another contractor backed into the temporary electric service.

3. Hamisch was observed by this inspector using the fire hydrant without a permit.

4. I seen that he was concealing the electrical work prior to an inspection.

5. After looking for blocked exit doors, the nightclub was checked for overcrowding.

6. Davies advised me that the concrete would be delivered at three o'clock.

7. Chapman was screaming and yelling curse words the whole time while I was writing my report.

8. We contacted individual residents of the building in there respective apartments.

9. Officer Pilak said, "I'am calling an ambulance and your going to be alright".

10. Superintendent Sherry Nowak said, "She heard two plumbers shouting at each other in the hospital parking lot and called police."

11. When I talked to she and her two children about Palmer's whereabouts they did'nt know where he was.

12. Everyone who attended the ceremony said they were very touched by the commissioners speech.

13. Here's the documents that need to go to Wendy and he in the front office.

14. A fire occurred at the Thompson's house while they were on there annual trip to a resort in the Bahama's.

15. The abuse was reported to Childrens' Services by a Social Worker at the High School.

16. Misuse of space heaters are an increasing problem in the elderly population.

17. Guidelines state that bleach and ammonia, which can be toxic are suppose to be stored in a locked cabinet or closet.

18. I contacted Fellowes for the purpose of issuing an invitation for him to make a presentation at our annual conference.

19. This officer pursued Carlson and upon encountering him; verbally commanded him to cease his attempt to kindle the rubbish fire.

20. I know their hopeing to return home tonight, however we have to wait until the firefighters tell us its safe to enter the area.

POST-TEST ANSWER KEY

Answers will vary.

1. Parker ~~tried to intimidate me~~ **threatened to punch me in the face** [or other words Parker said] when I told him I was writing a notice of violation.

2. ~~I asked Santos what happened, and he~~ Ramos said that another contractor backed into the temporary electric service. [Repetitious]

3. **I saw** Hamisch ~~was observed by this inspector~~ using the fire hydrant without a permit.

4. I ~~seen~~ saw that he was concealing the electrical work ~~prior to~~ **before** an inspection.

5. After ~~looking~~ **I looked** for blocked exit doors, ~~the nightclub was checked~~ **I checked the nightclub** for overcrowding. [dangling modifier, passive voice]

6. Davies ~~advised~~ **told** me that the concrete would be delivered at three o'clock. [Use *advised* only when you mean "counseled" or "recommended."]

7. Chapman was screaming ~~and yelling curse~~ words the whole time while I was writing my report. [A *curse* is an evil wish. Record word-for-word what Chapman said to you.]

8. We ~~contacted~~ **spoke** to ~~individual~~ residents of the building in ~~there respective~~ **their** apartments. [*Contacted* is vague: How did you get in touch with them? Omit *individual* and *respective*, which don't add any useful information. Be careful not to confuse *there* and *their*.)

9. Officer Pilak said, "~~I am~~ **I'm** calling an ambulance**,** and ~~your~~ **you're** going to be ~~alright".~~ **all right."** [Periods go *inside* quotation marks.]

10. Superintendent Sherry Nowak ~~said, "She~~ said she heard two plumbers shouting at each other in the hospital parking lot and called police. [Omit quotation marks when you're not quoting the person's exact words.]

11. When I talked to ~~she~~ **her** and her two children about Palmer's **whereabouts,** they ~~did'nt~~ **didn't** know where he was. [Comma Rule 1]

12. ~~Everyone~~ **All** who attended the ceremony said they were very touched by the **commissioner's** speech. [*Everyone* is singular and requires the clumsy phrase *he or she*. It's better to change the wording to *all*.]

13. ~~Here's~~ **Here are** the documents that need to go to Wendy and ~~he~~ **him** in the front office. [Think *documents are here* and *go to him*.]

14. A fire occurred at the ~~Thompson's~~ **Thompsons'** house while they were on ~~there~~ **their** annual trip to a resort in the **Bahamas**.

15. The abuse was reported to ~~Childrens'~~ Children's Services by a ~~Social Worker at the High School~~ social worker at the high school.

16. Misuse of space heaters ~~are~~ **is** an increasing problem in the elderly population. [Think **misuse is**]

17. Guidelines state that bleach and ammonia, which can be **toxic,** are **supposed** to be stored in a locked cabinet or closet. [Comma Rule 3]

18. I ~~contacted~~ **called** (or **emailed** or **wrote**) Fellowes ~~for the purpose of issuing an invitation for~~ **invite** him to make a presentation at our annual conference.

19. This officer ~~pursued~~ **chased** Carlson and **told** him to **stop trying to burn the rubbish**. ~~upon encountering him; verbally commanded him to cease his attempt to kindle the rubbish fire.~~

20. I know ~~their~~ **they're hoping** to return home tonight**. However,** ~~however~~ we have to wait until the firefighters tell us **it's** safe to enter the area. [You can also use a semicolon if you lower-case **however**.]

ANSWER KEY

Exercise 1 Why Are Reports Important?
page 16

Answers will vary. Here are some ideas you could have included in your letter:

- Your reports may be read by supervisors, the media, community leaders, and attorneys, who will be forming an opinion about you, based on what you've written.
- Your reports may help investigators uncover the truth about what happened.
- Accurate reports establish that you were properly following procedures.
- Effective reports provide information for statistical reports, help you prepare to testify in court, and may even keep you from having to testify.
- An effective report may help persuade the district attorney to prosecute a crime.
- A report may help investigators who are looking for a pattern of criminal behavior.
- You can improve your writing skills by studying the rules of English usage and asking a friend, relative, or co-worker to review what you've written.

Exercise 2 Rewrite a Paragraph
page 25

Here are some problems you might have noted and corrected:

- Labels like *Tenant 1* and *Tenant 2 1* don't add useful information and can be confusing later, especially if you're preparing to testify in court. Use people's names.
- Judgments and opinions ("acted like she didn't understand") do not belong in a report. Simply omit them.
- "trash was accumulated" is vague. A better description would include more details: "There were trash bags, construction materials, and miscellaneous clothes on the carport."

Exercise 3 Preparing to Write
page 31

1. Dealing with a victim's emotions
   a) is not part of an officer's job
   b) should usually be the first step in an interview CORRECT
   c) should be done only after all the facts are recorded
   d) is rarely necessary

2. "Chain of custody"
   a) refers to transporting a suspect
   b) refers to filing a report
   c) refers to evidence taken at the scene CORRECT
   d) does not need to be recorded in a report

3. Having extra paper and pens in a pocket
   a) may be helpful in an emergency CORRECT
   b) is unprofessional
   c) violates most agency's regulations
   d) may damage an officer's uniform

4. Which of the following does *not* need to be documented in a report?
   a) manufacturer and model of equipment
   b) address and legal description of the property
   c) Owner or owner's representative
   d) the inspector's theories about how the conditions occurred. CORRECT

5. Slang
   a) has no place in a report
   b) may require a definition if it's unfamiliar CORRECT
   c) should be used only if it's grammatical
   d) should be used only if it's easily understood

Exercise 4 Write a Report
page 39

Answers will vary. Here's how the report might be written:

At 1:30 p.m. I was dispatched to 11 Clover Lane for overcrowded conditions at the local Hindu Temple social hall. When I arrived, I saw a large number of people entering the building; some were carrying gifts. Blake Chardoury met me at the door and told me 100 guests were expected at his sister's wedding. He invited me in, pointed to the occupant load placard marked 150 persons, and led me out back, where chairs were set up for the ceremony.

I spoke to Deval Agarwal, a member of the board of trustees. He told me John Stem, who lives next door, has complained to him about the lack of parking and of temple guests blocking his driveway. Agarwal said they have made a concerted effort to make sure no one parks in front of his driveway by checking every 15 minutes before and during events. Agarwal said the issue may involve more than parking.

I went next door to speak with Mr. Stern. His driveway was empty, and no one answered the door. I left a card and returned to service after taking no action.

Exercise 5 Why Is the Report Necessary?
page 40

The report might be useful if John Stern has future complaints about the temple. You documented steps the temple has taken, and you made yourself available to discuss the problem.

In addition, if questions arise about how you handled the incident, you have shown that you took appropriate steps.

Exercise 6 Reports, Notices, Citations, Summonses, and Orders
page 44

Answers are in **bold type**.

1. Situation: Clothes dryers in an apartment building are being exhausted into the adjoining trash room.

IPMC Section 403.5 requires clothes dryer exhaust systems shall be exhausted outside the structure, or in accordance with manufacturer's instructions.

Order: Exhaust clothes dryers outside the building in accordance with IPMC 403.5.

2. Situation: The single required lighting fixture in a common apartment stairway is inoperable.

IPMC Section 402.2 requires stairways to be illuminated at all times.

Order: Repair or replace the light fixture in the stairway, in accordance with IPMC 402.2. The stairway must be illuminated at all times.

3. Thirty-six cardboard boxes of paper records are being stored in the exit stair tower of a bank building. IFC Section 315.2.2 states that combustible materials shall not be stored in exit enclosures.

Order: Remove all storage from the exit stair tower. IFC Section 315.2.2 prohibits storage in exit enclosures.

4. The manager of an apartment building is maintaining a Dumpster in the marked fire lane behind the building. IFC Section 503.4 prohibits obstruction of fire apparatus access roads.

Order: Remove the Dumpster from the fire lane. IFC Section 503.4 prohibits obstructions from fire apparatus access roads.

5. A homeowner has stopped maintaining the aboveground swimming pool in his back yard. The pool water is green and teeming with insects. IPMC Section 303.1 requires swimming pools to be maintained in a clean and sanitary condition.

Order: IPMC Section 303.1 requires swimming pools to be maintained in a clean and sanitary condition. Restore the pool to a clean sanitary condition, or remove it.

Exercise 7 Objectivity
page 49

Words and phrases that lack objectivity are printed in **bold type**.

√ 1. Mr. Bo applied for the permit to burn a two-foot square foot house as part of a Chinese funeral ceremony.

√ 2. The fire was part of a Hindu wedding ceremony.

X 3. The house was **filthy** and **reeked** of **stale** spices.

X 4. The fire was started by John Wells, who **appears to have mental problems**.

X 5. The grounds are **unkempt** and **unsightly**.

√ 6. From the sidewalk, I observed two junk cars in the back yard.

X 7. I red-tagged the furnace because **it posed a hazard**.

√ 8. The lumber stacks exceeded twenty feet in height and were leaning.

√ 9. The swimming pool water was green.

√ 10. I saw twelve cots and four air mattresses in the home.

Exercise 8 Rewrite These Sentences
page 53

Note: Details that you invented may be different from what is written here.

1. I ordered Robert Webster to extinguish the rubbish fire.

2. James Pate stammered when I asked if he had called for a close-in inspection. He said he thought the general contractor would call it in.

3. I arrived on the job site at 07:15 am and found that the sprinkler contractor had not arrived.

4. I ordered the homeowner to remove the bags of trash piled on his driveway.

5. I saw John Roberts lighting the controlled burn.

6. The electrical contractor said the general contractor would call in the inspection.

7. Mr. Wilson shouted "Get off my property, you SOB!" when I informed him that having an inoperative vehicle on his property was a violation of city code.

Exercise 9 What Did They Say?
page 57

√ 1. Mr. Jones said that the property is being used as a boarding house.

X 2. Jon Eastman **threatened me with violence** if I didn't get off his property.

X 3. The couple used **offensive language** at us when we got out of our city car.

X 4. Ms. Jones **disapproved** of the way the family was living.

√ 5. The blasting contractor applied for a permit on Tuesday, September 3, 2013.

X 6. Mrs. Jamison **kept complaining** about the dust from the blasting site.

√ 7. Mrs. Jamison said dust from the blasting site has forced her to run through the car wash all the time.

√ 8. Officer Ken Roberts told me the family has been sleeping in their car behind the shopping center.

Exercise 10 Think about OJT
page 61

Answers will vary.

Think about ways you could share what you've learned with new officers.

Exercise 11 Using Bullet Style
page 65

Note: These are suggestions only. Answers may vary. Notice that some information may not be suitable for bullet style.

1. I arrived at 3316 Fourth Street at 10:15 am and saw:

- the residents had evacuated and were assembled at the far edge of the parking lot

- many were in wheelchairs covered with blankets

- the fire department was already on the scene pumping water from a hydrant

- leaking water had frozen in the parking lot, causing a severe hazard.

I checked in with Chief Murphy at the command post. He asked me to:

- contact Public Works to request a sand truck

- contact the School Board to request school buses to temporarily shelter the residents

2. When I arrived at the building, I saw the following piled in the yard at the end of the driveway:

- two water closets

- two lavatories

- a stainless steel kitchen sink

- a plastic utility tub

A white utility van with "Bobby Fix-It" was parked in the driveway. No plumbing permit was posted. [This information doesn't match "I saw," so it shouldn't be formatted as bullets.]

3. After inspecting Sam's Great American Barbecue, I noted the following violations:

- the restaurant is without hot water

- there's an infestation of roaches in the kitchen and food storage area

- the refrigerator was 44 degrees.

- hot food on the buffet table had been maintained at 125 degrees for over four hours

I suspended the permit to operate. [This sentence doesn't match "I noted the following violations," so it shouldn't be formatted as bullets.

Exercise 12 Using Active and Passive Voice
page 69

Names in rewrites for sentences marked X will vary.

X 1. I saw Jones carrying plumbing fixtures into the building.

√ 2. Jones was carrying a toolbox and a pipe cutter.

X 3. Parker performed three hydrostatic tests on the automatic sprinkler system.

√ 4. Patterson was looking in his wallet for his contractor's license.

X 5. I questioned both witnesses.

√ 6. Finch hesitated and looked at his wife when I asked for his license.

√ 7. Chief Clancy and Major Hansen rewrote the procedure.

X 8. Chief Clancy and Major Hansen rewrote the policy two years ago.

√ 9. I was hoping to take a week of vacation in late August.

X 10. Keen found the wallet under the driver's seat.

√ 11. The mayor will be attending Lieutenant Cohen's retirement ceremony.

√ 12. Luis is interested in plan review.

X 13. Materials labs are paying their scientists top salaries right now.

√ 14. Three years ago, Luis was working in a low-paying service job.

_X 15. His boss said there wasn't much of a future for him there.

Exercise 13 Exploring Online Resources
page 73

Answers will vary.

Exercise 14 Fragments
page 80

√ 1. Williams Plumbing needs a final inspection for 9:30 this morning.

X 2. Needs to backfill the trench before tonight's rain.

X 3. Although, the work has been complete for three days.

X 4. Noticing that he has a pattern of requesting Friday inspections at the last minute.

√ 5. Other contactors have complained.

√ 6. They say it's unfair that Williams doesn't have to follow the rules.

√ 7. Williams uses impending weather or Monday holidays as an excuse.

X 8. Trying to manipulate the system.

X 9. Which creates disorder in the scheduling process.

√ 10. Inspector Link and I talked with Williams about scheduling inspections twenty-four hours in advance.

Exercise 15 Identifying and Correcting Run-on Sentences
page 83

1. Knudsen saw someone photographing the Rizzo house. No charges were filed.

2. When I entered the sun porch, I saw marijuana plants growing in front of the south window.

3. The emergency room was crowded. Duran signed herself out.

4. Culpepper said the suspect had a snake tattoo, gold hoop earrings, and two missing front teeth.

5. Carr insisted that because he was Belle's father, he could discipline her any way he chose.

6. I approached the dog. It growled at me.

7. Nieminen said she heard screeching brakes and a thud. She told her husband to go outside to look.

8. One car had a dented fender. The other was undamaged.

9. No one enjoys working holidays. However, in our profession it's often necessary.

10. I talked to the lieutenant. Then I went straight to the gym.

Here are the same sentences with semicolons:

1. Knudsen saw someone photographing the Rizzo house; no charges were filed.

3. The emergency room was crowded; Duran signed herself out.

6. I approached the dog; it growled at me.

7. Nieminen said she heard screeching brakes and a thud; she told her husband to go outside to look.

8. One car had a dented fender; the other was undamaged.

9. No one enjoys working holidays; however, in our profession it's often necessary.

10. I talked to the lieutenant; then I went straight to the gym.

Exercise 16 Misplaced Modifiers
page 86

Answers will vary.

X 1. While Peterson was holding the pipe wrench unsteadily in his right hand, it dropped off the scaffolding and struck the superintendent in his shoulder.

√ 2. We spotted the blaster's truck driving down Parker Avenue.

√ 3. After questioning Li, I left my card and asked him to call me if he recalled anything else about the deck collapse.

X 4. We saw parts of the circuit breaker panel that had shorted out Scattered around the room.

5. Inspector Pierarski found the little girl hiding behind a rosebush in the back yard.

Exercise 17 Parallelism
page 91

Answers may vary.

1. Connors told me she locked the door and turned on the alarm. She said a neighbor had the alarm code.

2. Ricky Lopez finished the basement without a permit, placed a construction Dumpster on the street without permission, and said the mayor wants him to bid on city contracts.

3. Each applicant must submit a birth certificate, take a physical examination, and come in for an interview.

4. In recent years we've been recruiting more women and minorities and taking a harder line on racism and sexism.

5. Always check your reports for accuracy, correct spelling, and completeness before you submit them.

Exercise 18 Using Semicolons
page 100

Answers will vary. Here are possible answers:

Luther Shalit is a math tutor in the prison GED program; he helps inmates learn elementary algebra and geometry.

He helps inmates learn elementary algebra and geometry; I've seen positive changes since he became a tutor.

I've seen positive changes since he became a tutor; Luther is proud of his knowledge and happy to be doing something useful.

Luther is proud of his knowledge and happy to be doing something useful; Luther has always been interested in mathematics.

Luther has always been interested in mathematics; before coming to prison he was planning to study bookkeeping.

> Captain Gephardt asked Linda Hammond to talk to us; she described her work as a Resource Officer at Penny Lane Middle School.
>
> She described her work as a Resource Officer at Penny Lane Middle School; she feels she's making a positive difference there.
>
> She feels she's making a positive difference there; discipline at the school has improved since she was assigned there.
>
> Discipline at the school has improved since she was assigned there; students trust her and come to her for advice.
>
> Students trust her and come to her for advice; she discusses substance abuse, family problems, and conflict resolution with students and faculty.

Exercise 19 More Practice with Semicolons
page 101

1. Vacant buildings are a big problem; the number has increased with the poor economy.

2. The office was quiet during the weekend although a few contractors called for inspections of emergency work.

3. I'm going to interview Davis this weekend; he may have some information about the building collapse.

4. I looked for the camera; however, it wasn't there.

5. We're looking for issues that might come up in the accreditation review, such as expired certifications and improper recordkeeping.

6. Our agency is planning a series of events to familiarize the community with our personnel and services.

7. The evaluation was a pleasant surprise; we received an excellent rating in several categories.

8. I found an open receptacle in the living room; Shipton found a cross-connection in the bathroom.

9. After the owner blocked the exit the second time, I issued him a summons.

10. The house is equipped with a silent intrusion alarm; furthermore, there are bars on the windows and doors.

Exercise 20 Comma Rule 1
page 106

1. While Inspector Josephs called for an ambulance, I disconnected the nail gun from the compressor.

2. The fight began when Todman insulted Jeffers.

3. Inspector Peters impressed the jury although he was nervous about testifying.

4. Because no shrubbery was broken, I knew the tenants didn't jump from an upstairs window.

5. If you talk to Wilson in the bedroom, I will interview his wife in the kitchen.

6. He has been in business since January 2004.

7. Because I suspected he wasn't licensed, I called the clerk's office.

8. We routinely check for business licenses after the permit has been issued.

9. When the fight broke out in apartment, Inspector Cary radioed for the police.

10. Although the surveillance camera wasn't working, the police found two eyewitnesses who saw the incident.

Exercise 21 Comma Rule 2
page 110

1. Bill Nelson stamped the plan and called the architect with the bad news.

2. I talked to Jerry Whitman, and Inspector Barthes questioned his wife.

3. Cashin produced a key but couldn't open the door.

4. The policy makes sense, but we can't implement it this year.

5. The treads and riser aren't uniform, and the handrail is too low.

6. The underground main failed both hydrostatic tests, and the thrust blocks are missing.

7. The shelter is overcrowded and does not provide enough services for displaced families.

8. I got out of my car and called for a police officer.

9. We questioned the subcontractor, but no one heard anything unusual that night.

10. I looked for rebar but didn't see any.

Exercise 22 Comma Rule 3
page 114

1. The plan tracking software that we purchased last month is already out of date.

2. Inspector Rice, who teaches in the academy part-time, has some good suggestions about preparing for the state certification exam.

3. During the winter, when many homeless people migrate to Florida, the crime rate increases here.

4. Our new permit counter, which opened last month, is better organized and more secure.

5. Abigail Gleason, who is the designer of record, visited the job site yesterday.

Exercise 23 Practice with Comma Rules 1, 2, and 3
page 115

1. As I approached the house, I heard a nail gun.

2. Linda grabbed her son's hand, and they ran down the street.

3. Linda grabbed her son's hand and ran down the street.

4. Paul, who just graduated from the code academy, is planning to go back for a degree.

5. I went back to Porter Street because I had more questions for Mrs. Smith.

6. Bailey's uniform, which should have been soiled, was suspiciously clean.

7. Menzies arrived at the meeting on time although traffic downtown was moving slowly.

8. Although the permit counter closed for repairs this week, it will reopen on Tuesday.

9. Glenn was afraid of confrontation at inspections at first but soon overcame his fears.

10. He spent extra time in the codebook and asked Inspector Kelly to work with him.

Exercise 24: Apostrophes
page 122

1. The sergeant's desk is cluttered with papers.

2. Her stepchildren's claims are unfounded.

3. The puppies were turned over to an animal shelter.

4. The puppies' condition is expected to improve.

5. Miss Jones' office is down the hall. [*Jones's* is also correct]

6. We all benefited from hearing James explain the new policy.

7. James' explanation cleared up several misunderstandings. [*James's* is also correct]

8. Families need to understand the special nature of police work.

9. Both instructors did an excellent job.

10. After a week's vacation, I was ready to return to work.

Exercise 25 More Practice with Apostrophes
page 123

1. I don't understand how to use this calculator.

2. Once again, the repairs to the Smiths' house didn't pass inspection.

3. Two days' work was lost when the computer system went down.

4. Lieutenant Conner asked me to address the family's concerns.

5. Last month's paychecks will be ready at nine o'clock.

Exercise 26 Using Quotation Marks
page 126

1. Katherine said that she had never spoken to her landlord, and she had mailed the rent check to Brooklyn.

2. Sarah told me, "I heard nail guns and knew they were working upstairs without a permit.'"

3. Brent said, "My electrician went to the supply house."

4. I asked, "Did you call for a final inspection?"

5. "I asked a neighbor if I could look at the house from her back yard," said Barton.

6. "The electrical system can't carry that air-conditioner," said Farrell.

7. Inspector Hoffman warned Rogers not to stock the store before his sprinkler final.

8. "When will you complete your code class?" I asked Susan.

9. Inspector West asked Linda Hamilton if she had heard any strange noises before the deck collapsed.

10. "Call an ambulance. Someone fell from the ladder," I shouted.

Exercise 27 Pronouns
page 130

Answers may vary.

1. Did everyone complete his or her requirements for inspector certification?

OR Did all the workers complete their requirements for inspector certification?

2. Its obvious that the academy needs to revise its curriculum.

3. Jill has more confidence taking tests than I [do].

4. Implementing the new policy is going to be difficult for the captain and me.

5. Everyone in the agency has been talking about his or her upcoming evaluations.

OR Everyone in the agency has been talking about the upcoming evaluations.

OR All the employees in the agency have been talking about their upcoming evaluations.

6. Lois replied to this email before she forwarded it to Mr. Morris and him.

7. Ken understands the procedure better than she [does].

8. The agency is proud of its safety record.

9. Someone didn't sign his or her timesheet for this month.

OR Someone didn't sign the timesheet for this month.

10. No one knows that part of town better than she [does].

Exercise 28 Verbs
page 137

1. We used to write all our reports by hand.

2. Inspector Larsen didn't see the memo about the new inspection procedure.

3. Pollard said she saw the contractor last Friday on the job site.

4. I sneaked Catherine a piece of candy during the meeting.

5. The notice of violation consists of three items and a correction order.

6. It's going to take a while for me to get used to the new administrative search warrant forms.

7. Perkins is supposed to be graduating from the code academy tomorrow.

8. Inmates from the vocational program did most of the interior work on the new administration building.

9. The report lists everyone who has a doctor's appointment today.

10. I've been studying so hard for this exam that I'm sure I'll pass with flying colors.

Exercise 29 Subject-Verb Agreement
page 141

Answers are in **bold** type; key words that help you find the correct answers are in *italics*.

1. *Twenty minutes* (**isn't**, aren't) long enough to fill out the form correctly.

2. *One* of the windows (**wasn't**, weren't) locked.

3. Neither the brakes nor the *clutch* (**seems**, seem) to be working properly.

4. Either the street valve is closed or *water main* (**is**, are) obstructed by debris.

5. *Advertising* for new positions (**is**, are) going to be posted tomorrow.

6. Departmental *policy* about interviews (**needs**, need) to be reviewed by an attorney.

7. *Each* of the subcontractors (**is**, are) telling us a slightly different story.

8. *All* of the subcontractors (is, **are**) in agreement on some of the details, however.

9. There (is, **are**) *problems* with Praeger's site plan.

10. There (**is**, are) a good *reason* why the Building Commissioner has doubts about this project.

Exercise 30 Capital Letters
page 145

1. My mother, father, and grandfather proudly attended my graduation from North Central Police Academy two years ago.

2. Although English and science have never been easy for me, I'm thinking of enrolling in college this fall.

3. The professors who teach building trades courses have an excellent reputation.

4. You'll enjoy taking Materials Science I and II with Professor Henry.

5. If you're not sure about a career, you should investigate the possibilities in engineering analysis and materials testing.

6. I'm seriously thinking about becoming an engineer, and my sister plans to become an architect.

7. I first became interested in code enforcement police work when an elementary school teacher taught our class about building inspections.

8. We meet weekly during the school year and did special projects in the summer when school was out.

9. Although Mr. West is retired, the staff at Tracy Elementary School still displays the safety posters he made.

10. Many inspectors received their first introduction to code enforcement working in the trades.

Exercise 31 Comparisons
page 148

Correct answers are in **bold** type.

1. I'd rather inspect a service station than work in the office.

2. Margaret is nearly as good at report writing as **he**.

3. Brock is the **better** of the two drivers.

4. Out of all the places I've worked, I like this agency **best**.

5. Few people face as many complaints as **we** in code enforcement.

6. Calvin is the **worst** driver in the department.

7. I'm good at setting up spreadsheets in Excel, and Gary knows almost as much as **I**.

8. I like outdoor work much more than sitting in an office cubicle.

9. I tried both laptops, and this one is definitely **better**.

10. Which of the three applicants is **most qualified**?

Exercise 32 Prepositions
page 153

Words that help determine the correct answer are formatted in *italics*.

1. In an apartment on the second floor, I found two cross-connections.

2. An *array* of arc fault circuit interrupters, ground fault circuit interrupters, and tamper-resistant receptacles was displayed on a table at our booth at the fair.

3. A *decision* about criminal charges for the unlicensed contractors is expected by tomorrow morning.

4. For now, the new commissioner is concentrating on getting to know the inspectors and the challenges they're facing. [You may also omit the comma because the prepositional phrase "for now" is so short.]

5. By the end of next October, we will be ready to discuss the new budget.

6. Under the sink in the kitchen, I found a leaking trap.

7. *Misuse* of temporary heaters by contractors *causes* significant property loss.

8. Of all the applicants for the new position, *Patricia Cooney seems* the most qualified.

9. An *assortment* of undersized fasteners *was* found when we investigated the porch collapse.

10. For the most part, unlicensed *contractors don't* seem to understand the seriousness of what they've done.

Exercise 33 Avoiding Common Errors
page 159 Corrections are in **bold type**.

1. When I **saw** smoke coming from the engine, I called the fire department.

2. The lock is **sticking. It** probably needs some graphite. OR The lock is **sticking; it** probably needs some graphite.

3. **A lot** of inspectors are watching the weather closely **because** the US Weather Service is predicting a hurricane.

4. I don't **know** the answer to that **question,** but **I** will find the answer for you.

5. Many people think that crime rates go up every **year,** but in many places crime rates have gone down.

6. Although I gave the data to **Mary,** Alice was the one who entered it into the database.

7. When I questioned Fischer about the **batteries**, he said he didn't know how they got there. [Omit quotation marks unless you're recording a person's exact words.]

8. Later Alan Jencks, Fischer's supervisor, said, "I don't allow any workers to remove batteries from the **shop.**"

9. We **did** everything we could to prepare our headquarters for the open house next Saturday. OR **We'd done** everything we could to prepare our headquarters for the open house next Saturday.

10. I'm signing up for a Spanish course next month **although** French might be more useful in some neighborhoods.

Exercise 34 Advised or Told?
page 175

1. I told Inspector Jones that he was assigned to the morning shift.

2. I advised Inspector Jones to improve his negative attitude.

3. I advised Mary Smith to see a doctor about the cut on her arm.

4. Smith told me that she cut her arm on the paper cutter.

5. Commissioner Simmons told us that he would be on vacation the first half of July.

6. Inspector Donaldson's doctor advised him to limit his cholesterol intake.

7. I already told the building manager about the broken alarm on the third floor.

8. I'm glad I listened to Mr. Johnson when he advised me to continue my education right after high school.

9. The chaplain told us that there would be a special religious service Sunday evening.

10. My guidance counselor in high school advised me to take a keyboarding course.

Exercise 35 Words Often Confused
page 184 Corrections are **in bold type**.

1. **We're** assigned to the committee that's drawing up a uniform policy about time off for work-related travel.

2. Not all agency **personnel** travel to meetings and conferences, so **it's** hard to be fair to everyone.

3. **You're** going to find that wearing **a** uniform doesn't automatically earn you respect.

4. If **it's all right** with the chief, we'll be replacing **a lot** of the exercise equipment upstairs.

5. Safety should always be our **principal** concern.

6. If the report is **too** long, I'll help you find places to cut it, and Mike can help us **too**.

7. Officer Campbell's master's in psychology is a nice **complement** to her undergraduate degree in law enforcement.

8. That car just **passed** me going sixty in a 30 mph zone.

9. The courtroom became **quiet** when Caruso went to the stand to testify.

10. **Who** is going to meet with the city council tomorrow?

Exercise 36 Use Plain English
page 190

Answers will vary. Here is one simplified version:

You're invited to a free Citizens' Code Academy on Saturday, November 14, from 9 to 4 in the Community Room at the Sixth Street Library. This is your opportunity to learn about Code Enforcement's safety programs and public initiatives. You can register by calling 555-1212 or clicking the link at www.SmithvilleCodes.org.

Exercise 37 Write Efficiently
page 190

utilize *use*

single-click *click*

PIN number *PIN*

for the purpose of *for*

the month of October *October*

yellow in color *yellow*

large in size *large*

in the event that *if*

if or when *if*

preplan *plan*

pull-down menu *menu*

take cognizance of *learn, discover, know*

lower down *lower*

scream and yell *scream (or yell)*

brand new *new*

this officer *I*

inquire *ask*

transport *drive*

pursue *chase*

abovementioned *above (or repeat the word or name)*

commence *begin*

relative to *about*

initiate *begin*

in order to *to*

residence *home, house, condo, apartment*

ascertain *discover, find out*

numerous *many*

at the present time *now (or omit)*

being that *because*

finalize *end*

endeavor *try*

by means of *by*

in close proximity to *near*

modify *change*

with reference to *about*

terminate *end, finish*

affirmative *yes, nodded*

approximately *about*

Exercise 38 Understanding Public Records
page 196

Answers will vary.

INDEX

A
Active voice 60, 67-8
Advised/told 174-5, 174
Agreement, pronoun 127-9
Agreement, subject-verb 135, 139-42, 151-2
Apostrophes 119-24, 129, 133, 157

B
Bias 23, 47-50
Brevity 52-3, 61
Bullet style 38, 60, 63-6
   headings 64

C
Capitalization 143-6
Chain of custody 24, 30
Ciarlone case 14-15
Citations 44
Commas 103-116, 152, 157-8, 163
   chart 95
   comma rule 1 103, 105-8, 152, 157-8
   comma rule 2 99, 100, 103, 109-12, 159, 161, 162
   comma rule 3 104, 113-14, 162-3
   overview 103-4
Common errors 155-60
Comparisons 130, 147-50, 181, 183
Completeness 30, 91
Confusing words 177-86
Coordinate conjunctions 100, 109, 159, 161, 162

D
Dangling modifiers 85-88
Disposition 36, 38-39

E
Efficiency 24, 64
Emotions 29

F
FANBOYS words 100, 109
First person 22, 60, 129, 162
Fragments 77-80
Freedom of Information Act 170, 193-6

G
Grammar myths 161-4

H
Hearsay 28-9

I
I, me 22, 23, 60, 129, 162
Interviews 29
Interrupters 113-4, 162-3
It's, its 129, 179-80

J
Jargon 52, 71, 171-5, 187-92

M
Misplaced modifiers 85-88

N
Names 23-4, 37, 42, 119, 144, 162-3
Narrative 15, 19, 20, 33, 35, 37-8, 42, 59
Notice of violation 19, 27, 42-3

INDEX

O
Objectivity 19, 47-50, 56
Online resources 71-3
Orders 44
Organizing the report 33-4

P
Parallelism 89-92
Passive voice 51-2, 60, 67-70
Periods 97-102, 125, 155, 156, 158, 163
Plain English 187-92
Post-test 197-8
Post-test Answer Key 199-201
Prepositions 151-4
Prepositional phrases
    with commas 152
    with pronouns 140-1
    with verbs 140, 151-2
Pretest 5-7
Pretest Answer Key 8-10
Probable cause 14
Pronouns 60, 127-32
    agreement 127-8
    case 129, 152, 156-7
    in comparisons 130, 147-8
Punctuation 97-126
    apostrophes 119-24, 133, 157
    commas 103-16
    periods 97-102
    quotation marks 29, 55-8, 125-6, 160
    semicolons 97-102

Q
Quotations 53, 55-8, 115-6

R
Reports
    opening sentence 33, 35, 42
    organizing 33-4
    planning 27-32
    samples 20-1
    stages of writing 22
Run-on sentences 81-4, 155

S
Semicolons 97-102
Sensitivity 49
Sentences 95-116
    sentence pattern chart 95
Slang 24-25, 30, 49, 56, 162, 168-9
    curse/swear 133, 179
Spelling 25, 91, 121, 136, 158
Station Nightclub fire, 14
Subject-verb agreement 139-42
Subordinate conjunctions 105-8, 157-8
Summonses 41, 168

T
Taking notes 27
Terminology 167-70
Texting 143, 159
"Thumb rule" 129, 152, 156-7

V
Verbs 133-8, 161
    agreement 139-42
    with helpers 133, 159, 182
    with prepositional phrases 140-1, 152

W
Who/Whom 184
Words to avoid 49, 61, 171-6

About the Authors

About Jean Reynolds

Dr. Jean Reynolds is Professor Emerita at Polk State College in Winter Haven, Florida, where she taught English for over 30 years. She holds a doctorate in English from the University of South Florida and has published nine books, including *The Criminal Justice Report Writing Guide for Officers* and (with the late Mary Mariani) *Police Talk*. She has taught advanced report writing, instructor techniques, and communication skills at the Kenneth C. Thompson Institute of Public Safety at Polk State College.

About David Diamantes

David Diamantes is a fire protection consultant, code trainer, curriculum developer, and author. He served with the Fairfax County Fire and Rescue Department in Virginia for 25 years as a firefighter, company officer, and fire prevention officer. He retired in 1998 and to work as a consultant and trainer. He is the author of two textbooks, *Fire Prevention: Inspection and Code Enforcement* and *Principles of Fire Prevention* (Delmar-Cengage), and he was a contributing author for the current edition of *Building Department Administration*, published by the ICC. He's a certified fire inspector, fire investigator, plans examiner, and residential building inspector.

www.ingramcontent.com/pod-product-compliance
Lightning Source LLC
Chambersburg PA
CBHW080240180526
45167CB00006B/2357